GW00578053

The Geometry of Evolution

The metaphor of the adaptive landscape – that evolution via the process of natural selection can be visualized as a journey across adaptive hills and valleys, mountains and ravines – permeates both evolutionary biology and the philosophy of science. The focus of this book is to demonstrate to the reader that the adaptive landscape concept can be put into actual analytical practice through the usage of theoretical morphospaces – geometric spaces of both existent and nonexistent biological form – and to demonstrate the power of the adaptive landscape concept in understanding the process of evolution. The adaptive landscape concept further allows us to take a spatial approach to the concepts of natural selection, evolutionary constraint and evolutionary development. For that reason, this book relies heavily on spatial graphics to convey the concepts developed within these pages, and less so on formal mathematics.

GEORGE McGHEE is Professor of Paleobiology in the Department of Geological Sciences at Rutgers University, New Jersey, USA. He is a past Fellow of the Konrad Lorenz Institute for Evolution and Cognition Research in Vienna, Austria.

The Geometry of Evolution

Adaptive Landscapes and Theoretical Morphospaces

GEORGE R. McGHEE JR.
Rutgers University

CAMBRIDGE UNIVERSITY PRESS
Cambridge, New York, Melbourne, Madrid, Cape Town,
Singapore, São Paulo

Cambridge University Press
The Edinburgh Building, Cambridge CB2 2RU, UK
Published in the United States of America by Cambridge
University Press, New York

www.cambridge.org
Information on this title: www.cambridge.org/9780521849425

First published 2007

Printed in the United Kingdom at the University Press, Cambridge

A catalogue record for this publication is available from the British Library

Library of Congress Cataloguing-in-Publication Data

McGhee, George R.
The Geometry of evolution : adaptive landscapes and theoretical
morphospaces / George R. McGhee, Jr.
p. cm.
Includes bibliographical references and index.
ISBN-13: 978-0-521-84942-5 (hardback)
ISBN-10: 0-521-84942-X (hardback)
1. Evolution (Biology)–Mathematical models. 2. Adaptation
(Biology)–Mathematical models. 3. Morphology–Mathematical models.
I. Title.

QH371.3.M37M34 2006
576.801′5118–dc22 2006026495

ISBN-13 978-0-521-84942-5 hardback
ISBN-10 0-521-84942-X hardback

For Marae
A leannain m'òige, mo bhean.

Contents

Preface

The concept of the adaptive landscape is the creation of the great American geneticist Sewall Wright who, along with the equally great British scientists R. A. Fisher and J. B. S. Haldane, crafted the Neo-Darwinian synthesis of evolutionary theory in the 1930s. The metaphor of the adaptive landscape, that evolution via the process of natural selection could be visualized as a journey across adaptive hills and valleys, mountains and ravines, permeated both evolutionary biology and the philosophy of science through the succeeding years of the twentieth century. Yet critics of the adaptive landscape concept have maintained that the concept is of heuristic value only; that is, it is fine for creating conceptual models, but that you cannot actually use the concept in analysing the evolution of actual animals or plants. That criticism became invalid in the year 1966 when the palaeontologist David M. Raup used computer simulations to model hypothetical life forms that have never existed in the evolution of life on Earth, and who subsequently created the concept of the theoretical morphospace.

The focus of this book is to demonstrate to the reader the power of the adaptive landscape concept in understanding the process of evolution, and to demonstrate that the adaptive landscape concept can be put into actual analytical practice through the usage of theoretical morphospaces. The adaptive landscape concept allows us to visualize the possible effects of natural selection through simple spatial relationships, rather than complicated modelling of changing environmental or ecological conditions. For that reason, this book relies heavily on spatial graphics to convey the concepts developed within these pages, and less so on formal mathematics.

I thank the Santa Fe Institute for the invitation to visit and work on computational methods in theoretical morphology in 2000, for it was at the Santa Fe Institute that the idea of writing this book came to me in conversations with Dave Raup. I thank the Konrad Lorenz Institute for Evolution and Cognition Research for the Fellowship that enabled me to work at the institute in 2005, for it was there that I developed many of the ideas presented in Chapters 7 and 8 of this book. Finally, I thank my wife, Marae, for her patient love.

1

The concept of the adaptive landscape

> The idea of a fitness landscape was introduced by Sewall Wright (1932) and it has become a standard imagination prosthesis for evolutionary theorists. It has proven its worth in literally thousands of applications, including many outside evolutionary theory.
>
> *Dennett (1996, p. 190)*

What is an adaptive landscape?

An adaptive landscape is a very simple – but powerful – way of visualizing the evolution of life in terms of the geometry of spatial relationships, namely the spatial relationships one finds in a landscape. Consider an imaginary landscape in which you see mountains of high elevation in one region, towering mountains separated by deep valleys with precipitous slopes. In another region these mountains give way to lower elevation rolling hills separated by wide, gently sloping valleys, and that these further give way to broad flat plains in the distance. Now replace the concept of 'elevation' (height above sea level) with 'degree of adaptation' and you have an adaptive landscape. Why is that such a powerful concept? The purpose of this book is to answer that question.

The concept of the adaptive landscape was first proposed by the geneticist Sewall Wright in 1932. Being a geneticist, he thought in terms of genes rather than morphology and Darwinian fitness rather than adaptation, and his original concept is what is termed a fitness landscape today, rather than an adaptive landscape. The two concepts differ only in that the dimensions of a fitness landscape are genetic traits and

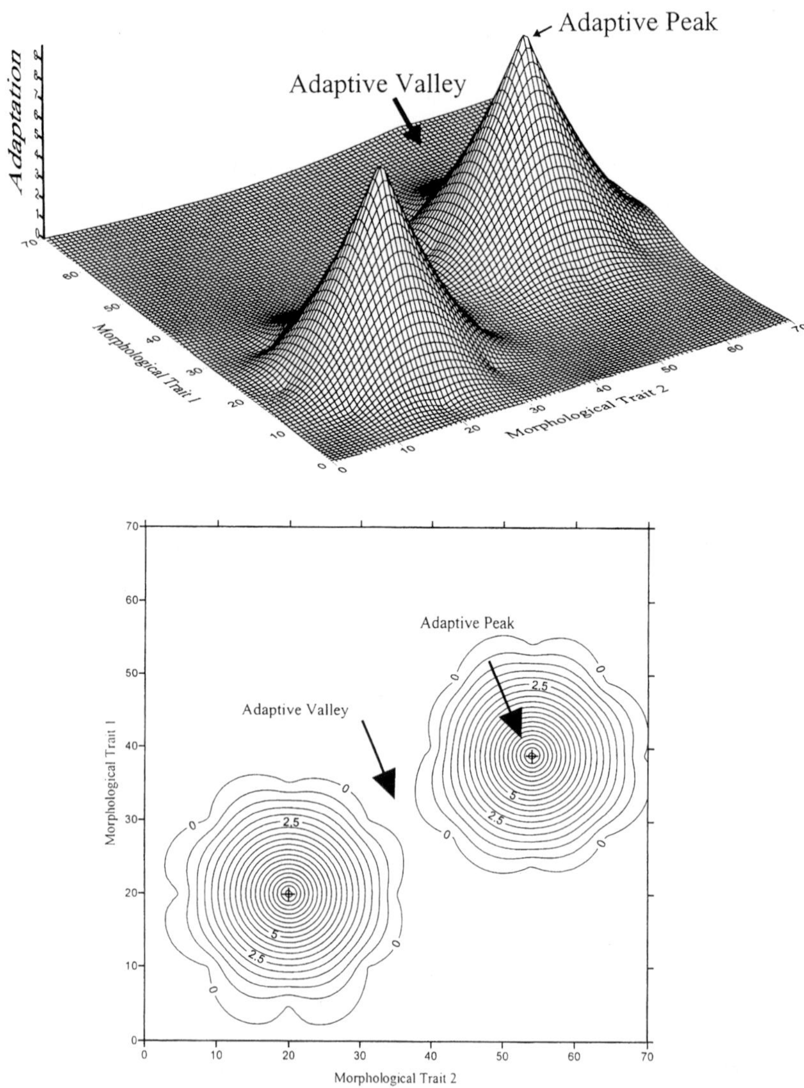

Figure 1.1. A hypothetical adaptive landscape, portrayed as a three-dimensional grid at the top of the figure and a two-dimensional contour map at the bottom. Topographic highs represent adaptive morphologies that function well in natural environments (and therefore are selected for), and topographic lows represent nonadaptive morphologies that function poorly in natural environments (and therefore are selected against). In the contour map portrayal, the top of an adaptive peak is indicated by a plus-sign, following the convention of Sewall Wright (1932).
Source: Modified from McGhee (1980a).

degree of fitness whereas the dimensions of an adaptive landscape are morphological traits and degree of adaptation (Fig. 1.1). A fitness landscape is used by geneticists to visualize evolution, and an adaptive landscape is used by morphologists. As I am a morphologist, a student of the evolution of biological form, this book will concentrate on adaptive landscapes and, beginning in Chapter 4, the very important related concept of the *theoretical morphospace*.

But back to Sewall Wright. His first crucial insight was that it could be possible (at least theoretically) to construct a space of all possible genetic combinations that living organisms might produce, and that one could visualize such a complex space by simply considering the possible combinations of two genes at a time or, in the case of an adaptive landscape, two morphological traits at a time (Fig. 1.1). That is, if genetic trait number one had 10 different variants or alleles, and genetic trait number two had 10 different variants, then the total possible genetic combinations of those two traits would be 100 potential variants.

Wright's second crucial insight was that *the majority of those 100 possible variants probably do not exist in nature*. Perhaps only 10 of the possible variants actually exist as living organisms, the other 90 variants potentially could exist but do not. Why not? Wright proposed that these 90 potential genetic combinations had zero fitness; that is, they represented lethal genetic combinations. The other 10 variants had fitness values greater than zero, some perhaps having higher fitness than others. Wright further proposed that these genetic relationships could be spatially visualized as geometric relationships by simply adding the dimension 'degree of fitness' to the two genetic trait dimensions, producing a three-dimensional grid similar to a landscape (as in the adaptive landscape in Fig. 1.1). If the landscape is portrayed in two-dimensions by using fitness contours to give the fitness dimension, then the result looks very much like a topographic map of a landscape. The 10 existent combinations of genetic traits number one and two would be located on the peaks or slopes of the hills within the landscape (depending upon their degree of fitness), and the 90 possible but nonexistent combinations of genetic traits number one and two would be located in the flat plain of zero fitness. Thus was born Sewall Wright's concept of the fitness landscape.

In adaptive landscapes the *high regions* are called adaptive peaks, and the *low regions* between the peaks are called adaptive valleys (Fig. 1.1). The degree of adaptation of the possible morphological traits is determined by functional analyses of the potential forms; that is, analyses of how well the potential morphological variants function in nature. The geometric arrangement of the adaptive peaks within the landscape thus represents, in a spatial fashion, the different possible ways of life available to organisms. The spatial distribution of the adaptive valleys and plains represents ecomorphologies that are nonfunctional in nature.

Modelling evolution in adaptive landscapes

Adaptive landscapes are potentially very powerful tools for the analysis of the evolution of life. Life is constantly evolving, and we would like to know why life has evolved the way that it has in the past three and one-half thousand million years of Earth history, and perhaps be able to predict how life might evolve in the future.

Although evolution itself is a fact, an empirical observation, the cause of evolution is theoretical. That is, there exist several different theories to explain how evolution takes place. The most widely subscribed-to theory of how evolution takes place is that of natural selection, first proposed by Charles Darwin. If he had not proposed it, Alfred Wallace would have instead; thus it was clearly an idea whose time had come in the 1800s. What is natural selection? A precise, rather pithy definition is the 'differential change in genotypic frequencies with time, due to the differential reproductive success of their phenotypes' (modified from Wilson and Bossert, 1971). The first part of the definition ('differential change in genotypic frequencies with time') is simply a restatement of evolution itself, in that evolution is genetic change in populations from generation to generation. The real heart of the theory is 'differential reproductive success' of various phenotypes, or morphologies. If certain organisms with certain morphologies in a population reproduce at a higher rate than other organisms with other morphologies, then the next generation will contain more of their genes than the previous one. And that change in gene frequencies, from generation one to generation two, is by definition evolution. Thus natural selection could clearly drive evolution.

The definition of natural selection does not specify what causes differential reproductive success; it simply holds that if it does occur, evolution will result. The next question is obvious: what determines the differential reproductive success of differing phenotypes, or morphologies, such that different animals and plants reproduce at different rates? It is here that the concept of adaptation enters the equation. Organisms must function in their environments, and they must interact with other organisms. If some organisms possess morphologies and behaviours (aspects of their phenotypes) that allow them to function well in their ecological setting then they are described as well adapted. Well adapted organisms are healthy, well fed and potentially able to devote more time and energy to reproduction. If other organisms possess morphologies that do not allow them to function as well – say, they cannot run as fast due to the different structure of their legs, or cannot find their prey or other food as quickly due to the different structure of their eyes or ears (their visual and auditory systems) – then they are described as poorly adapted. Poorly adapted organisms must spend more time simply trying to escape predators and to find food, are generally less healthy and spend less time and energy in reproduction.

Wright's concept of a fitness, or adaptive, landscape is firmly rooted in the theory of natural selection (we shall see in Chapter 4 that the concept of the theoretical morphospace *is not*). In the previous section we have seen that an adaptive landscape is an actual spatial map of the different possible ecomorphologies that are available to organisms, and of other possible ecomorphologies that are nonfunctional and thus not available to organisms. What would happen now if we place a population of actual organisms within the adaptive landscape, say half-way up the side of an adaptive peak, and observe the evolution of that population with time? A basic rule of modelling evolution in adaptive landscapes is that *natural selection will operate to move a population up the slope of an adaptive peak, from lower degrees of adaptation to higher degrees of adaptation.* That seemingly simple rule has some intriguingly complicated consequences, however, as we shall see in the next chapter.

2
Modelling natural selection in adaptive landscapes

> 'Wedges in the economy of nature' wrote Darwin in his diary, leaving us with a glimpse of his own first glimpse of natural selection ... Later biologists, by the fourth decade of the twentieth century, would invent the image of an adaptive landscape whose peaks represent the highly fit forms, and see evolution as the struggle of populations of organisms driven by mutation, recombination, and selection, to climb toward those high peaks. Life is a high-country adventure.
>
> *Kauffman (1995, p. 149)*

Visualizing natural selection

We have seen in the last chapter that an adaptive landscape is a way of visualizing the evolution of life in terms of the geometry of the spatial relationships one finds in a landscape, where the landscape consists of adaptive hills and valleys. If we use the theory of natural selection to model evolution within an adaptive landscape, we saw that natural selection will operate to move a population up the slope of an adaptive peak, from lower degrees of adaptation to higher degrees of adaptation.

What happens, however, when an evolving population reaches the top of an adaptive peak? Or what happens if an evolving population encounters two peaks in an adaptive landscape, rather than one? Clearly natural selection will operate in different ways at different times in the evolution of any group of organisms, depending upon the environmental and ecological context within which that group of organisms is evolving. The adaptive landscape concept allows us to visualize the possible effects of natural selection through simple spatial relationships, rather than complicated modelling of changing environmental or ecological conditions.

Modelling directional selection

Let us consider again the situation where a population of animals of plants is positioned half-way up the slope of an adaptive peak. In this situation, natural selection will operate to move the population up the slope of an adaptive peak, from lower degrees of adaptation to higher degrees of adaptation. But how does natural selection actually accomplish this?

Natural selection operates on variation in nature. If there were no variation in nature, natural selection would cease (note that evolution itself may not cease, however, because evolution may be driven by more than natural selection – we shall consider this possibility in more detail later). That is, if a population of animals is composed of individuals that are all identical in the state of their adaptive morphologies – for example, if they are all clones inhabiting the same environment – then they should all function equally as well in a given environment. Natural selection would have no differences in adaptive morphology to 'select' and all the individuals should reproduce at more or less the same rate, with some random variation. Such a hypothetical situation is very rare in nature, however, where variation is the normal natural condition.

There are two main sources of variation for natural selection to operate with. One is genetic recombination, the other is genetic mutation. Genetic recombination is the constant reshuffling of genes that occurs from generation to generation in sexually reproducing organisms. Imagine all the genes present within a species, its genome, to be represented by a deck of 52 playing cards, where each card is a gene. Imagine further that the morphology of any individual animal is determined by four cards (genes), then you can divide your deck of 52 cards into 13 individual animals in generation number one. Each time the animals reproduce to produce a new generation you reshuffle all the cards again and draw another 13 sets of four cards each. In this simple exercise (actual genetic recombination is much more complicated) you can easily see how much variation is produced from generation to generation by merely reshuffling the same genes over and over again. Genetic mutation, on the other hand, is the appearance of a new genetic coding – a new card in the deck that was not present there previously.

Consider that our hypothetical population is composed of two major variants, animals with A-type morphologies and animals with B-type morphologies, and that in generation number one the population is equally divided in numbers of individuals with A-type and B-type morphologies (Fig. 2.1). However, let us further imagine that animals with A-type morphologies function a bit better in the environment than animals with B-types; that is, A-types have a somewhat higher degree

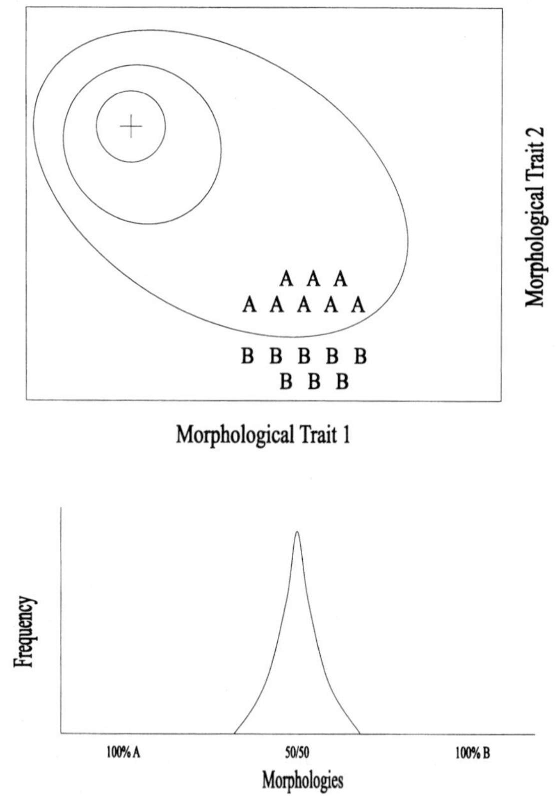

Figure 2.1. Modelling directional selection, part one. The spatial positions of individuals of a hypothetical species population, composed of organisms with morphological variants A and B, are depicted within an adaptive landscape in the top figure. Individuals with morphological variants A have a higher adaptive value than B (these variants are shown upslope from B) and thus, under the expectations of the theory of natural selection, organisms with A-type morphologies should reproduce at a higher rate than those with B-type morphologies. The initial frequency of organisms with morphological variants A and B within the species is depicted as roughly equal in the graph given in the bottom figure.

of adaptation than B. We can represent this selective difference in an adaptive landscape by bisecting the population with an adaptive contour, where individuals with A-type morphologies are on the upslope side of the contour, and individuals with B-type morphologies are on the downslope side (Fig. 2.1).

Now let us consider the state of the population after several generations of reproduction under the influence of natural selection. We would predict that the somewhat better adapted animals with A-type morphologies should reproduce at a somewhat higher success rate than the less well-adapted animals with B-type morphologies, and that the number of individuals with A-type morphologies now comprise a greater percentage of the total population numbers than individuals of B-type (Fig. 2.2). That is, we now have more individuals in the population on the upslope side of the adaptive gradient than on the downslope side (Fig. 2.2). The population is moving uphill.

Sooner or later, however, the uphill movement of the population will cease when all of the individuals in the population have A-type morphologies, and all have the same degree of adaptation. Now we need to introduce the second source of variation into the equation: genetic mutation. Let us introduce three new morphological variants into the scenario at random, as genetic mutation is random. One new variant, X, has a morphology that is further upslope than the parent population; another new variant, Y, has a morphology that is further downslope than the parent population; the last variant, B, is a backmutation to a previously existent morphology (Fig. 2.3). Under the expectations of the theory of natural selection, variants Y and B will be selected against – they will have an even lower success rate of reproduction than the individuals in the parent population that produced them. On the other hand, however, variant X will be selected for, and would have a higher success rate of reproduction than the individuals in the parent population itself. With time, individuals with X-type morphologies should become more and more numerous, and individuals with A-type morphologies less numerous; thus the population continues to move uphill.

The effect of natural selection in this particular scenario is termed *directional selection*. We can summarize the effects of directional selection in Figure 2.4 with a series of vectors that indicate that the effect of natural selection will always be to select genetic mutational morphologies that possess higher degrees of adaptation, and that the net result of natural selection is evolution that always proceeds in the uphill direction in an adaptive landscape.

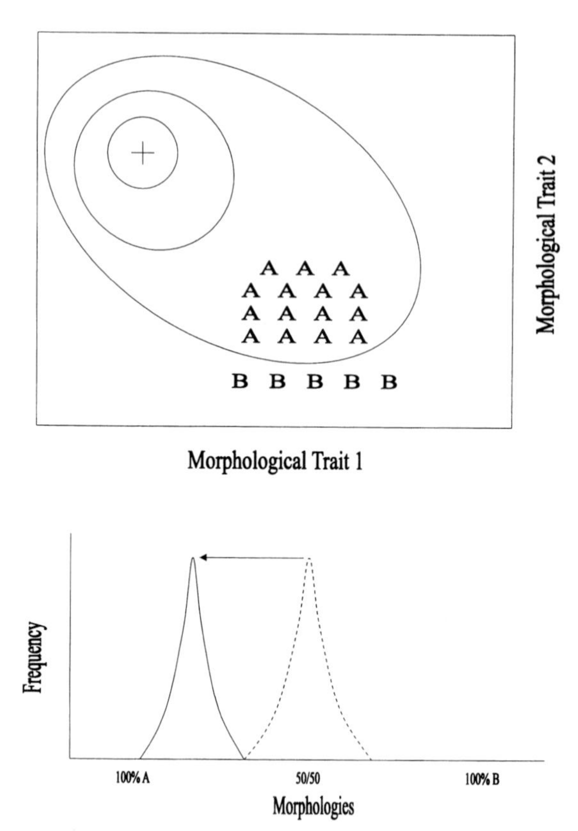

Figure 2.2. Modelling directional selection, part two. The spatial positions of individuals of the hypothetical species population within the adaptive landscape after several generations of natural selection (top figure). The number of organisms possessing the higher-adaptive A-type morphologies has increased within the species (top figure), and the frequency distribution of morphologies within the species has shifted to the left in the graph given in the bottom figure.

Modelling stabilizing selection

As directional selection operates to produce evolution in the uphill direction, sooner or later the evolving animals or plants will reach the adaptive peak, the local point of maximum degree of adaptation within the adaptive landscape. What happens then?

Once at the peak, any major new source of variation will always be in the downslope direction, and thus will be selected against. Consider a hypothetical population sitting on top of an adaptive peak, and

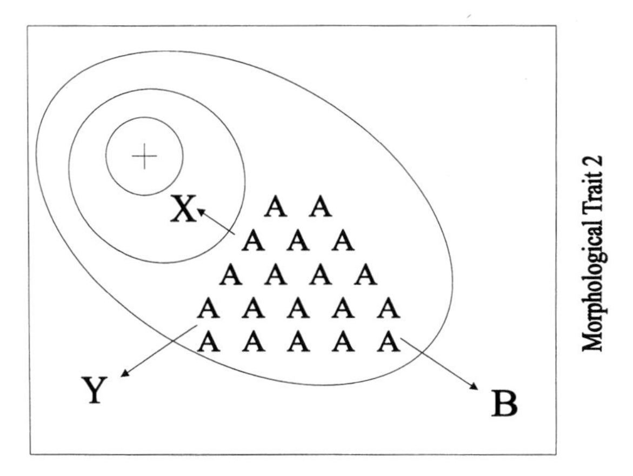

Figure 2.3. Modelling directional selection, part three. The effect of new genetic mutation in an adaptive landscape. After several generations of natural selection, the hypothetical species population now is composed of organisms that all possess morphology A. Three new morphological variants arise randomly by mutation: one new mutation is upslope (X) and two new variants are downslope (Y, and the backmutation B). Under the expectations of the theory of natural selection, organisms with morphology X should reproduce not only at a higher rate than organisms with morphologies Y and B, but should also reproduce at a higher rate than than those of the parent population itself (organisms with A-type morphologies).

three new random mutational variants with morphologies X, Y and Z (Fig. 2.5). If the variants are different, they must of necessity occur outside the boundaries of the pre-existing morphologies already present in the population. However, regardless of which new morphology is randomly produced, the variants will always be in the downslope direction if the parent population is already sitting on top of an adaptive peak. The new variants will be less well adapted than individuals in the parent population and will be selected against.

The effect of natural selection in this particular scenario is termed *stabilizing selection*. We can summarize the effects of stabilizing selection in Figure 2.6 with a series of vectors that indicate that the effect of natural selection will always be to select against genetic mutational morphologies that deviate from the maximally adapted morphologies already present in the parent population; that the net result of natural selection is to return the population to its original state and the cessation of further evolutionary changes in morphology.

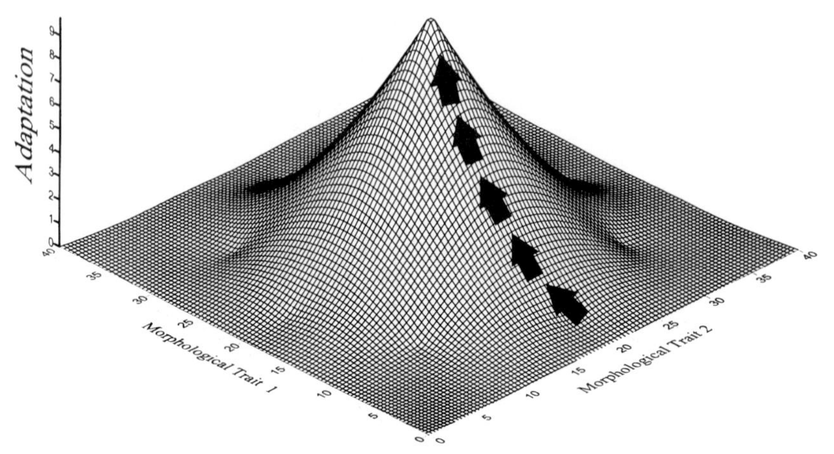

Figure 2.4. Modelling directional selection, part four. The vectors summarize the effect of natural selection in an adaptive landscape. Natural selection should always favour genetic mutational morphologies that possess higher degrees of adaptation; thus evolution should always proceed in the upslope direction within the landscape.

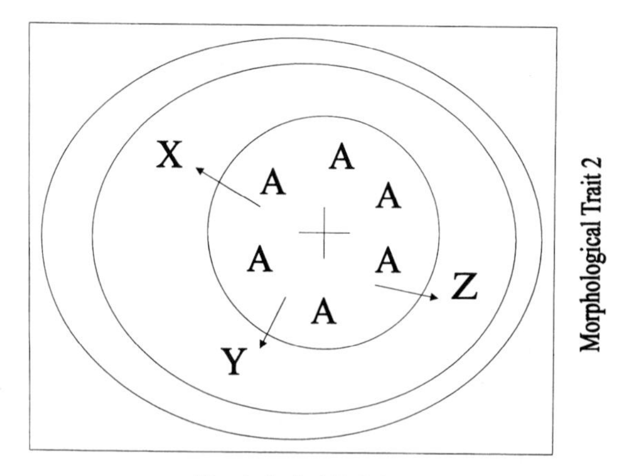

Figure 2.5. Modelling stabilizing selection, part one. The spatial positions of individuals of a hypothetical species population, all of which possess A-type morphologies located at the top of an adaptive peak (all located within the highest adaptive contour shown in the figure). Three new morphological variants arise randomly by mutation (X, Y and Z), all of which have a lower adaptive value than the morphologies of the parent population (the variants are all in the downslope direction) because the parent population morphology occupies the top of the adaptive peak. As the new morphological variants are less well adapted than individuals in the parent population, they will be selected against.

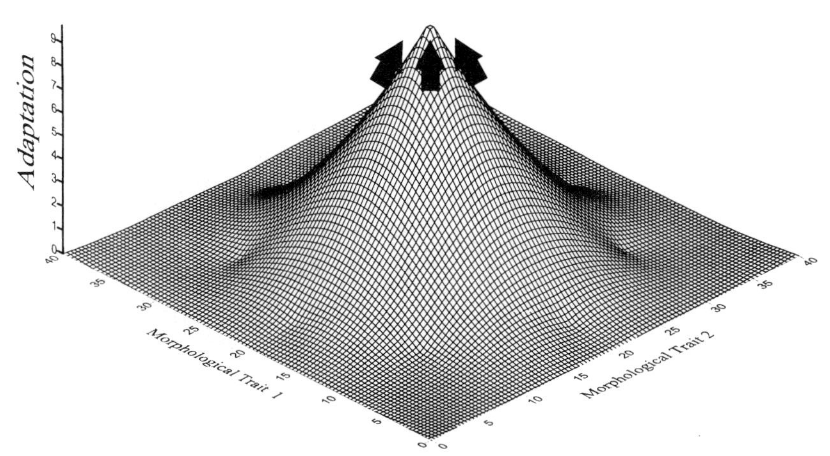

Figure 2.6. Modelling stabilizing selection, part two. The vectors summarize the effect of natural selection on a species population occupying the top of an adaptive peak. All new genetic mutational morphologies that deviate from the maximally adapted morphologies of the parent population will be selected against, resulting in the cessation of further evolutionary changes in morphology.

Modelling disruptive selection

Environments are variable in nature, just as animal and plant morphologies are. In some parts of the Earth you may stand on a craggy, rocky mountainside and view a broad, green glen only a few hundred metres away in one direction, and see the ocean itself only a few hundred metres away in another direction.

Thus it is entirely conceivable that a single species population may encounter a variety of different environments, each with different selective properties in terms of which morphologies function well in those environments, and which do not. This possibility is easy to model by simply adding additional adaptive peaks to the adaptive landscape, where each peak represents a different environment and selective condition.

Consider the case of a population that finds itself positioned mid-way between two adaptive peaks in an adaptive landscape. Imagine there are three variants in the population: animals with morphological types A, B and C, where animals with B-type morphologies are the most abundant in the population (Fig. 2.7). However, let us further imagine that animals with A-type morphologies function a bit better in the environment characterized by adaptive peak number one, that animals

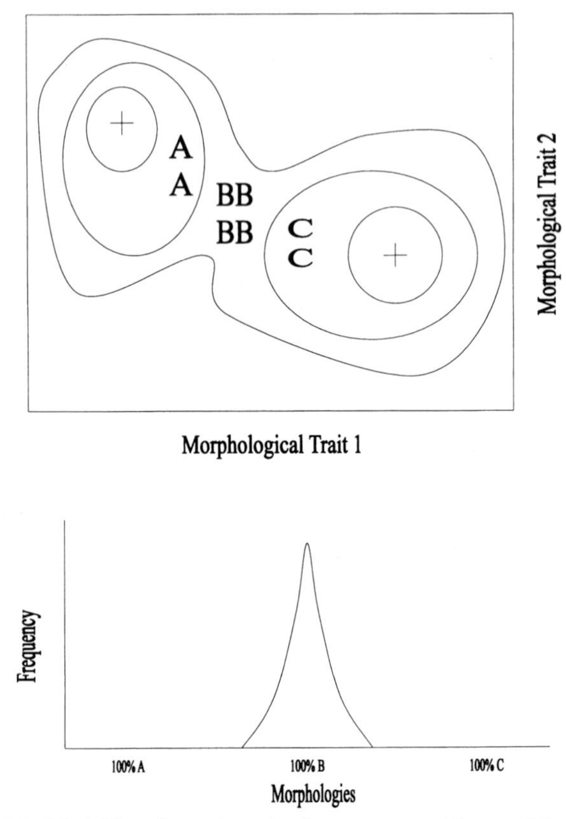

Figure 2.7. Modelling disruptive selection, part one. The spatial positions of individuals of a hypothetical species population, composed of organisms with morphological variants A, B and C, are depicted within an adaptive landscape (top figure). Organisms with A-type and C-type morphologies have higher adaptive values than B (A-type morphologies are upslope from B towards the peak to the left, and C-type morphologies are upslope from B towards the peak to the right) and thus, under the expectations of the theory of natural selection, organisms with morphological variants A and C should reproduce at a higher rate than those with B. The initial frequency of organisms with morphological variants A, B and C within the species is depicted in the graph given in the bottom figure, where the B-type morphology is shown to be the most abundant and the frequency distribution tails off in the A-type and C-type directions.

with C-type morphologies function a bit better in the environment characterized by adaptive peak number two, and that than animals with B-type morphologies do not function quite as well in either environment. We can represent this selective difference in an adaptive landscape by trisecting the population with two adaptive contours, where individuals with A-type morphologies are on the upslope side of the contour in the direction of the peak on the left in the landscape, individuals with C-type morphologies are on the upslope side of the contour in the direction the peak on the right and individuals with B-type morphologies are on the downslope sides of both contours in the valley between the peaks (Fig. 2.7).

Now let us consider the state of the population after several generations of reproduction under the influence of natural selection. We would predict that the somewhat better adapted animals with morphological types A and C should reproduce at a somewhat higher success rates than the less well-adapted animals with B-type morphologies, and that the number of individuals with morphological types A and C now comprise a greater percentage of the total population numbers than individuals of B-type (Fig. 2.8). That is, we now have more individuals in the population on the upslope side of the adaptive gradients in the direction of both peaks one and two than on the downslope side in the valley between the peaks (Fig. 2.8). The population is splitting apart.

As we modelled before, let us introduce new morphological variants into the scenario at random through genetic mutation. One new variant, X, has a morphology that is further upslope than individuals with A-type morphologies, and another new variant, Y, has a morphology that is further downslope (Fig. 2.9). Likewise, one new variant, P, has a morphology that is further upslope than individuals with C-type morphologies, and another new variant, Q, has a morphology that is further downslope (Fig. 2.9). Under the expectations of the theory of natural selection, variants Y and Q will be selected against – they will have an even lower success rate of reproduction than the individuals of morphological types A and C in the parent population that produced them. On the other hand, however, variants X and P will be selected for, and will have a higher success rate of reproduction than the individuals of morphological types A and C in the parent population. With time, individuals with morphology X and P should become more and more numerous, and individuals with morphological types A, B and C less numerous, thus the end-points of the original population continue to move uphill and the population continues to split apart.

The effect of natural selection in this particular scenario is termed *disruptive selection*. We can summarize the effects of disruptive selection in Figure 2.10 with two vector trails that separately climb two different adaptive peaks. The net result of disruptive selection is the splitting of an

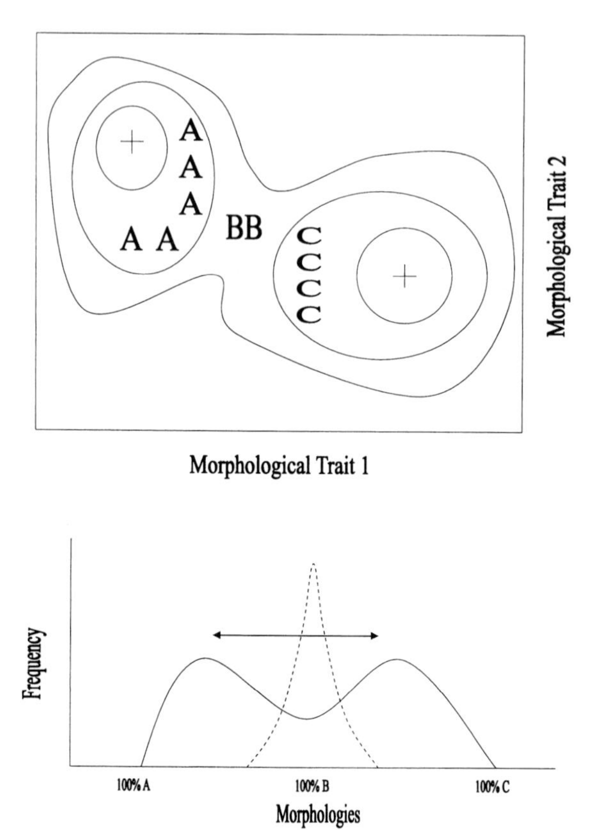

Figure 2.8. Modelling disruptive selection, part two. The spatial positions of individuals in the hypothetical species population within the adaptive landscape after several generations of natural selection (top figure). The number of organisms possessing morphological types A and C has increased within the population, whereas the number of organisms possessing B has decreased, leading to the shift in the frequency distribution of morphologies depicted in the graph given in the bottom figure. Thus the initial unimodal frequency distribution of morphologies (Fig. 2.7) has now become bimodal. The population is splitting apart.

ancestral species population into two (or more) descendant species populations with different adaptive morphologies.

Modelling less-than-optimum evolution

Thus far we have seen that quite different evolutionary scenarios result, depending upon the location of a population within an adaptive

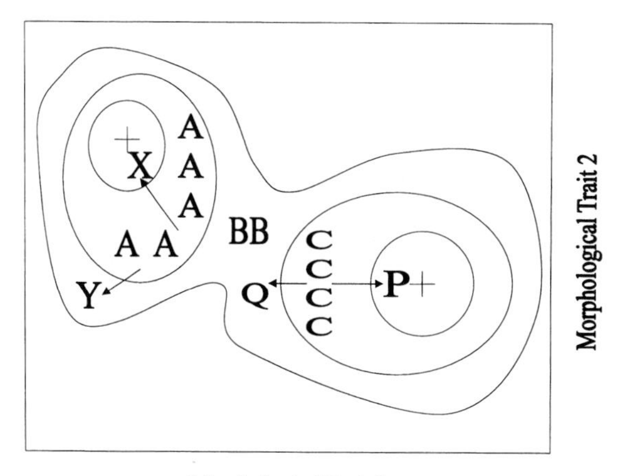

Figure 2.9. Modelling disruptive selection, part three. The effect of new genetic mutation in an adaptive landscape. Four new morphological variants arise randomly by mutation: one new mutation is upslope on the peak to the left (X), and one new mutation is upslope on the peak to the right (P). Two new variants are downslope, one down from the peak to the left (Y) and one down from the peak to the right (Q), and have the same adaptive value as the ancestral morphology B. Under the expectations of the theory of natural selection, organisms with new morphologies X and P should reproduce at a higher rate than organisms with morphological variants A and C in the parent population. The new morphologies Y and Q, and the organisms with ancestral morphology B, will not be favoured by selection. Thus with time, individuals with morphologies X and P should become more and more numerous, and individuals with ancestral morphological variants A, B and C less numerous, thus the end-points of the original population continue to move uphill and the population continues to split apart.

landscape, from the application of the same modelling rule that natural selection operates to move a population up the slope of an adaptive peak, from lower degrees of adaptation to higher degrees of adaptation. Now let us envision a situation where the strict application of this rule results in a population winding up with a less than optimum morphology through the action of natural selection!

Consider the adaptive landscape illustrated in Figure 2.11. A high adaptive peak exists with a smaller adaptive peak located on the slope of the taller peak. An evolving population has climbed the adaptive slope to the top of the smaller peak, the location of the local adaptive maximum. However, the higher adaptive position of the taller peak is

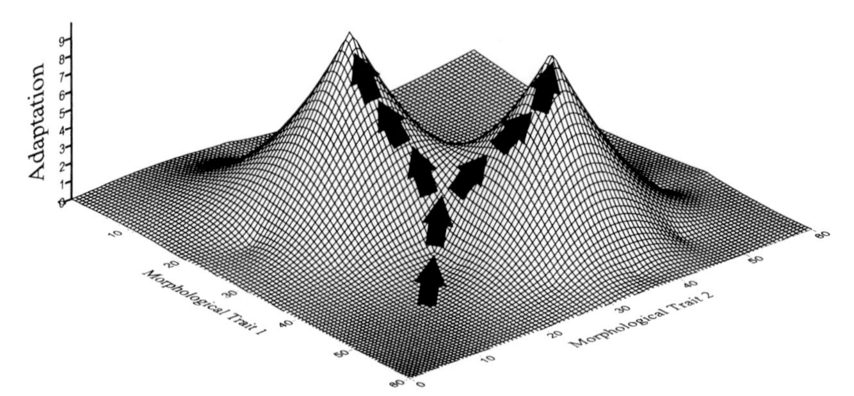

Figure 2.10. Modelling disruptive selection, part four. The vectors summarize the effect of disruptive natural selection in an adaptive landscape. Natural selection should always favour genetic mutational morphologies that possess higher degrees of adaptation, thus evolution should always proceed in upslope directions within the landscape, resulting in two vectors trails that separately climb two different adaptive peaks. The net result of disruptive selection is the splitting of an ancestral species population into two (or more) descendant species populations with different adaptive morphologies.

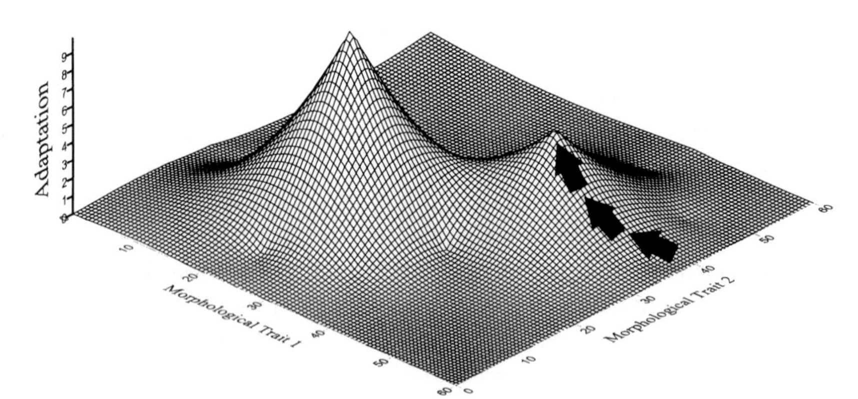

Figure 2.11. Modelling less-than-optimum evolution. A species population evolving under the influence of natural selection can only explore the local adaptive possibilities. The evolving population depicted by the vectors in the figure has climbed the local smaller adaptive peak and is now separated by an adaptive valley from the much higher adaptive peak located to the left. Stabilizing selection will now act to keep the population confined to the smaller peak, with a less-than-optimum adaptive value.

within sight, so to speak, so surely the population will continue to evolve to eventually conquer it, yes?

The answer is no. *A population evolving under the influence of natural selection can only explore the local adaptive possibilities.* That is, the higher peak is not 'in sight' at all to the population stuck on the local adaptive maximum, where the action of stabilizing selection will operate to keep it positioned.

Evolutionary topology of adaptive landscapes

In the previous example we have seen that it is possible to construct more and more complicated evolutionary scenarios in an adaptive landscape simply by adding additional adaptive peaks. Now let us consider the possible *shapes and arrangements* of those peaks. If you go hiking in the mountains, you immediately notice that not all mountains are alike. Some are very high, with precipitous slopes. Others are lower, and have more gently rounded slopes. Just as in a real landscape, the peaks and hills in an adaptive landscape may come in different sizes, shapes, and arrangements.

The theoretician Stuart Kauffman (1993, 1995) has conducted extensive computer simulations of evolution via the process of natural selection in what he calls '*NK* fitness landscapes'. In *NK* fitness landscape models, *N* is the number of genes under consideration and *K* is the number of other genes which affect each of the *N* genes. The fitness of any one of the *N* genes is thus a function of its own state plus the states of the *K* other genes which affect it, allowing one to model epistatic genetic interactions. Such interactions can be extremely complex, yet still can be modelled with the computer.

Kauffman's computer simulations have demonstrated that two end-member landscapes exist in a spectrum of *NK* fitness landscapes: a 'Fujiyama' landscape at *K* equal to zero, and a totally random landscape at *K* equal to *N* minus one, which is the maximum possible value of *K*. In the Fujiyama landscape a single adaptive peak with a very high fitness value exists, with smooth slopes of fitness falling away from this single peak (Fig. 2.12). Such a fitness landscape exists where there are no epistatic interactions between genes, where each gene is independent of all other genes. At the other extreme, every gene is affected by every other gene, and a totally random fitness landscape results, a landscape comprised of numerous adaptive peaks all with very low fitness values.

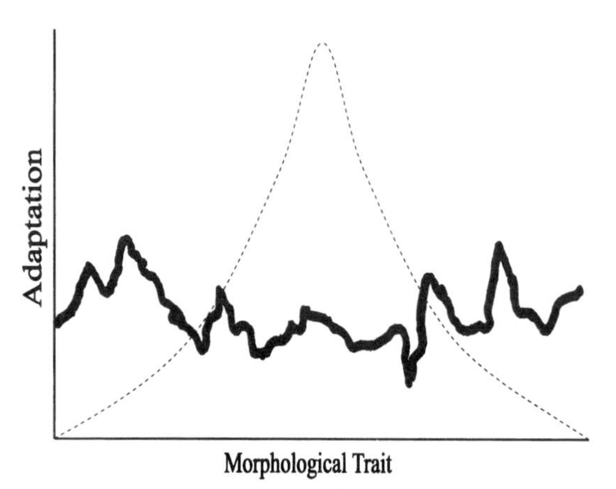

Figure 2.12. Contrasting topologies of adaptive landscapes. The dashed line depicts a 'Fujiyama' landscape, with a single adaptive peak with very high adaptive value, versus the solid line depicting a 'rugged' landscape, with multiple peaks of varying height but all of much lower adaptive value than the Fujiyama peak.

In a Fujiyama landscape a single adaptive maximum occurs, in a random landscape any area in the landscape is just about the same as any other area. Between these two extremes exists a spectrum of landscapes, ranging from 'smooth' (a few large peaks) to increasingly 'rugged' (multiple smaller peaks; Fig. 2.12), and from 'isotropic' (landscapes where the large peaks are distributed uniformly across the landscape; Fig. 2.13) to 'nonisotropic' (landscapes where the large peaks tend to cluster together; Fig. 2.13).

Kauffman (1993, 1995) has argued that the process of evolution on Earth appears to have taken place on rugged fitness landscapes and not on Fujiyama landscapes, smooth landscapes of high peaks that he characterizes as the Darwinian gradualist ideal. Computer simulations of the process of evolution via natural selection in rugged fitness landscapes reveals on the one hand that the rate of adaptive improvement slows exponentially as the evolving population climbs an adaptive peak, but on the other hand that the highest peaks in the landscape can be climbed from the greatest number of regions! The latter conclusion is in accord with the empirical observation that the phenomenon of *convergent morphological evolution* has been extremely common in the evolution of life on Earth, a phenomenon that we shall examine in more detail in the next chapter.

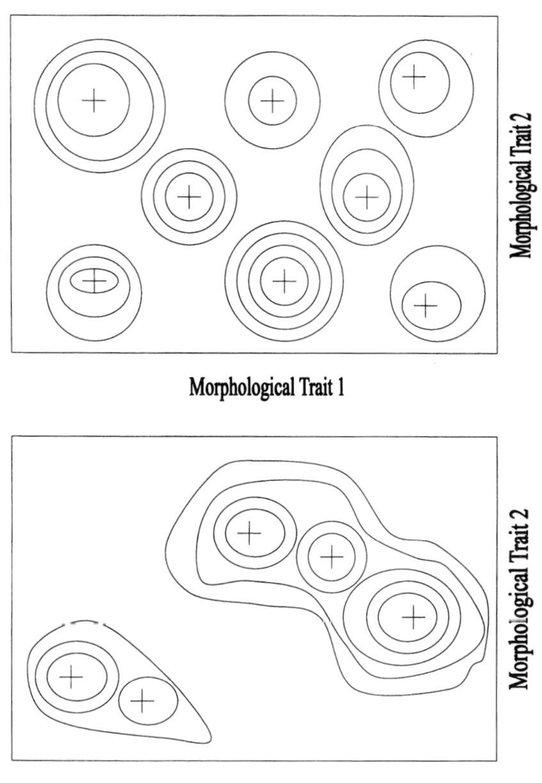

Figure 2.13. Contrasting topologies of adaptive landscapes. The top figure depicts an 'isotropic' landscape, where the adaptive peaks are uniformly distributed across the landscape. The bottom figure depicts a 'nonisotropic' landscape, where the adaptive peaks cluster near one another in several groups.

If life has evolved on rugged fitness landscapes then epistatic interactions must be the norm, and the fitnesses of morphological character states must be correlated. Kauffman (1993, 1995) has argued that the more interconnected the genes are the more conflicting constraints arise. These conflicting constraints produce the multipeaked nature of the rugged landscape (Fig. 2.12). There exists no single superb solution as in a Fujiyama landscape. The conflicting constraints of the intercorrelated genes produce large numbers of compromise, less than optimum, solutions instead. A rugged landscape results, a landscape with numerous local peaks with lower altitudes.

Kauffman's computer simulations are based on models of genetic interactions and their consequences. However, we can take his fitness landscapes and transform them to adaptive landscapes by simply changing their dimensions (as we saw in Chapter 1), and use them to explore the consequences of the different adaptive landscape topologies for the evolution of morphology. In doing so, we must keep in mind the caveat of Arnold et al. (2001) that a complex fitness landscape of genotypes does not automatically produce a corresponding complex adaptive landscape of phenotypes. We shall be exploring the morphological effects of geometry, not genetics.

Why has evolution not ceased?

Regardless of the shape and arrangement of adaptive peaks, natural selection will operate to move a population up the slope of an adaptive peak, from lower degrees of adaptation to higher degrees of adaptation. Sooner or later, every evolving animal or plant (if evolving via natural selection) should wind up on top of a local adaptive peak. Once a species reaches the top of an adaptive peak, stabilizing selection should operate to keep that species in that position in the adaptive landscape. Evolution should cease.

Evolution has clearly not ceased (we, the species *Homo sapiens*, are a mere 200,000 years old). Yet we know that life has existed on Earth for at least three and one-half thousand million years, and that surely should have been enough time for all life to have reached all possible adaptive peaks, or not? Why does life continue to evolve new forms?

If we examine this question using the concept of the adaptive landscape we can quickly see that there are two possible causes of continued evolution: the first is the possibility that life might be able to overcome the effect of stabilizing selection by jumping directly from one adaptive peak to another without going downslope into the adaptive valley between them. Evolution via jumping from one peak to another is an interesting concept, the consequences of which the adaptive landscape concept allows us to quickly visualize. In Figure 2.14 we see a variety of adaptive peaks, some low, some high. The higher the peak, the deeper the valley produced by their longer adaptive slopes. The distance within the landscape between the higher peaks is thus greater than the distance between the lower peaks, and the depth of the valley between the higher peaks is deeper than the depth between the lower peaks. A jump from

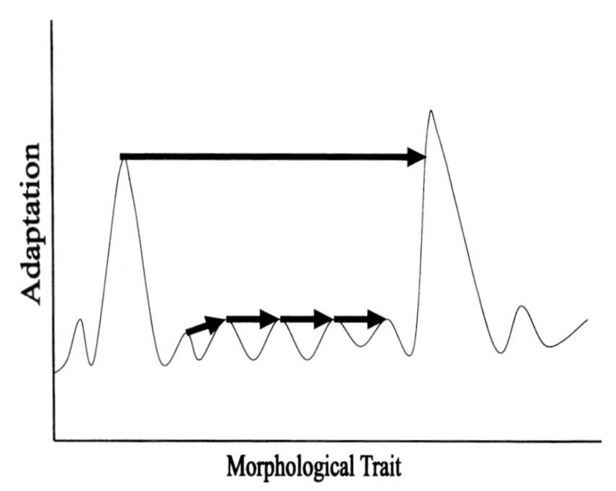

Figure 2.14. Modelling evolution via peak jumping in an adaptive landscape. The higher the peak, the deeper the valley between the peaks. A jump off a high peak is much more likely to result in a large drop in adaptive value of the new mutant morphology than jumps off low peaks surrounded by shallow adaptive valleys. In such a landscape, random short jumps off low peaks are much more likely to be successful than random long jumps off high peaks. Thus the adaptive landscape concept predicts that evolution by peak jumping should occur in organisms that are not highly adapted, in organisms that are generalists in their environments rather than highly adapted specialists.

one high peak to another high peak thus requires a long distance to be covered, whereas the closer proximity of the smaller peaks requires only a short jump to go from one to another.

In such a landscape (Fig. 2.14) we can predict that random short jumps off low peaks are more likely to reach an adjacent peak successfully than random long jumps off high peaks. Such a conclusion matches the empirical observation of geneticists that larger random mutations are more likely to be lethal than smaller mutations. That is, if you make a long jump off a tall peak in the landscape, the most probable consequence is a drop in the degree of adaptation, a long fall into the valley. Thus the adaptive landscape concept would predict that, if evolution by peak jumping occurs, then it should occur in organisms that are not highly adapted, in organisms that are generalists in their environments rather than highly adapted specialists.

The second possibility is that the *adaptive peaks themselves are not stable in time*. The morphology that has a high degree of adaptation today may not have that same high degree of adaptation tomorrow;

that is, the position of the adaptive peak in the landscape has moved, or the peak itself has vanished entirely, leaving behind only a flat nonadaptive plain.

The position of an adaptive peak in the landscape is a function of environmental and ecological factors, of abiotic and biotic conditions. Clearly, those conditions may change with time. In fact, the longer the period of time that elapses, the less likely that the environmental and ecological conditions that were present at the beginning of that period of time will still be present at the end of that period of time. What is now fertile farmland and forest in parts of northern Europe and North America was frozen tundra just eleven to twelve thousand years ago, and the varied habitats of most of present day Canada, Norway and Sweden did not exist at all, as those regions were covered by immense ice caps. Times change.

Modelling environmental and ecological change

Environmental and ecological change can be either gradual or abrupt, continuous or quantal. We can model continuous, gradual environmental change in an adaptive landscape by allowing the adaptive peaks to move their positions within the landscape. We can model abrupt, quantal environmental change in an adaptive landscape by allowing the adaptive peaks to collapse, or to drastically lower their altitudes (the degree of adaptation), and then to return to their previous position and altitude when the environmental disturbance has passed.

The major question in modelling gradual environmental change is: how fast is the environment changing? That is, how fast are the adaptive peaks moving across the landscape (Fig. 2.15). In our previous modelling, evolving organisms climb stationary adaptive peaks under the influence of natural selection. Now organisms must not only climb the peaks, but must also follow the peaks as they shift their positions in the landscape. Thus a great deal of the continuity of evolution through geological time may be modelled with unstable adaptive landscapes, landscapes that change with time (Snoad and Nilsson, 2003).

The speed at which various organisms can evolve to follow shifting adaptive peaks is a function of their variability and mutation rate. Obviously, if organisms cannot evolve rapidly enough to follow the rate at which peaks are moving in the landscape, they will actually *move downslope with time as the peak moves out from under their position on*

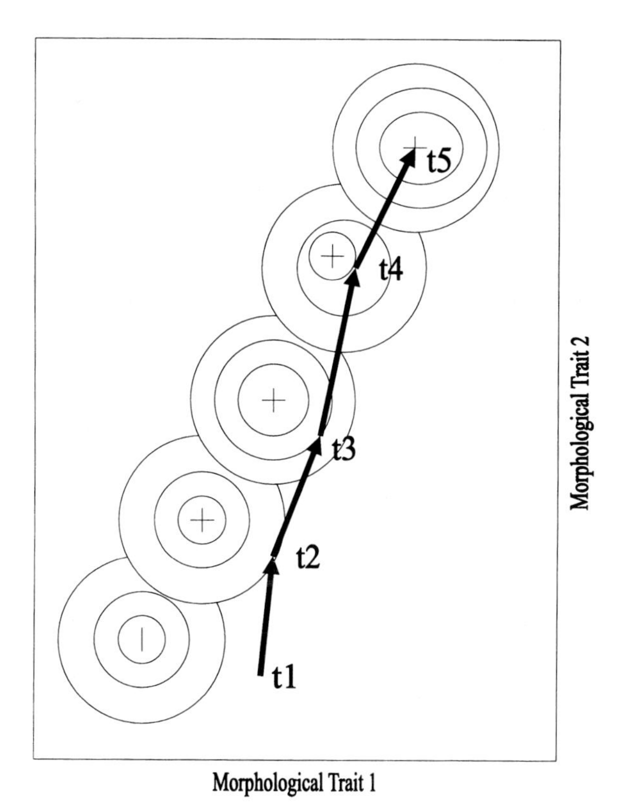

Figure 2.15. Modelling evolution via shifting adaptive peaks in an adaptive landscape. Now the modelled species populations must not only climb adaptive peaks, they must follow the peaks as they shift their positions in the adaptive landscape. The illustrated hypothetical species population exists on a low adaptive position at time = t1, but progressively climbs to higher and higher adaptive contours in positions time = t2 through time = t5, while simultaneously exhibiting large changes in morphology simply as a result of evolving to keep pace with the shifting adaptive peaks across the landscape (the track of the vectors).

the landscape. Thus we can use the concept of the adaptive landscape to also model the deterioration of the degree of adaptation of an animal or plant morphology in the face of environmental change, and not just its improvement. In this latter case, natural selection still favours those variants with the higher degree of adaptation, but the organisms simply cannot evolve better adaptations fast enough to match the changing environmental conditions. Indeed, if the peak moves so fast as to entirely outpace the rate of evolution of the organisms upon it, then the organisms eventually are left behind on the plane of zero adaptation and become extinct (Fig. 2.16). Thus we can easily model the process of extinction (or one of the processes, as there are more than one) with the adaptive landscape concept (for models of extinction on Stuart Kauffman's *NK* fitness landscapes, see Solé, 2002; Newman and Palmer, 2003).

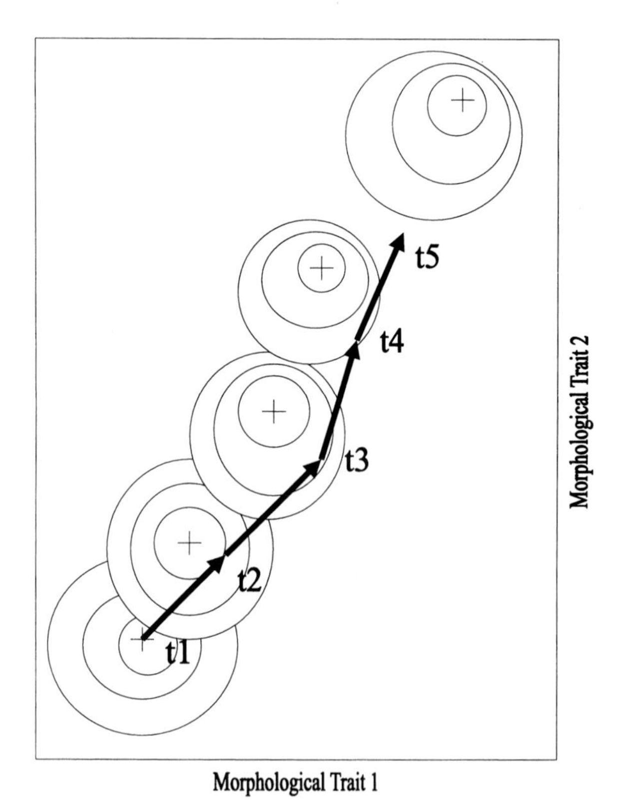

Figure 2.16. Modelling extinction via shifting adaptive peaks in an adaptive landscape. In this scenario, the hypothetical species populations cannot evolve fast enough to keep pace with the adaptive demands of changing environmental or ecological conditions, modelled as the shifting of adaptive peaks across the adaptive landscape. The illustrated hypothetical species actually moves downslope with time, from its adaptive value at position time = t1 to a lower adaptive contour at position time = t2 and an even lower adaptive contour at time = t3. At position time = t5 the adaptive peak has moved entirely out from under the species (its morphologies are now entirely maladaptive) and the species goes extinct.

There exists a very interesting scenario midway between the two extremes of species successfully evolving to follow moving adaptive peaks (Fig. 2.15), and species unsuccessfully keeping up, falling behind and becoming extinct (Fig. 2.16). This is the case where species can only evolve fast enough to exactly maintain their same adaptive position on a moving adaptive peak (Fig. 2.17). That is, they evolve fast enough not to slide downslope with time, yet they cannot evolve fast enough to climb upslope on the peak. They are constantly evolving, but perpetually stuck at the same degree of adaptation.

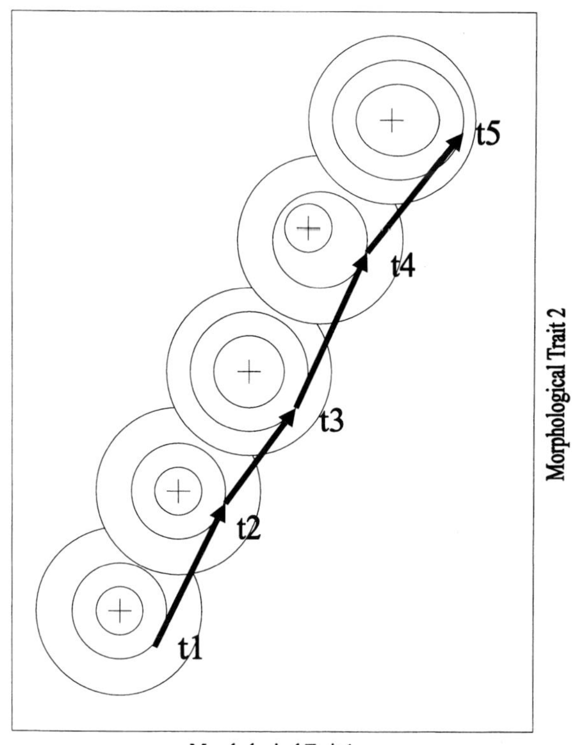

Figure 2.17. Modelling the Red Queen Hypothesis in an adaptive landscape. In this intermediate scenario between those modelled in Figs. 2.15 and 2.16, the hypothesized species evolves fast enough to remain on a shifting adaptive peak, but not fast enough to climb the peak to higher levels of adaptation. Thus the species is constantly evolving new morphologies but perpetually stuck at the same degree of adaptation (same contour level in the landscape), in essence 'constantly running in order to remain in the same place' like the Red Queen in Lewis Carroll's *Through the Looking-Glass*.

This intriguing possibility is called the *Red Queen Hypothesis* of evolution, after the Red Queen in Lewis Carroll's *Through the Looking-Glass* who told Alice that she had to constantly keep running just in order to stay in the same place (Van Valen, 1973). It is a rather bleak evolutionary possibility, in that no adaptive improvement ever takes place, and the probability of a species going extinct in such a world is always a constant regardless of whether the species is young, in existence only a few hundred years, or if the species has been present on the Earth for millions of years.

Moving on from gradual change, the major question in modelling quantal environmental change is: how severe is the environmental disturbance? If the environmental disruption is severe but not lethal, we can model the disturbance by lowing the altitude (the degree of adaptation) of the adaptive peak. That is, the morphologies that normally function very well do not now function nearly as well during this time of environmental or ecological disruption. For example, small song-birds that normally fly very well will have great difficulty flying when a powerful hurricane or cyclone is passing through their habitat. But if the birds take shelter and wait for the passing of the storm, then the normal adaptive benefits of their flying abilities will return when the environmetal disruption has ceased. The adaptive peak has returned to its former altitude.

If, however, the environmental disruption is lethal then we can model it by collapsing the adaptive peak entirely (Fig. 2.18). That is, where once there existed a high altitude peak in the adaptive landscape there now exists only a flat plain of zero adaptation. Morphologies that used to work very well now have no adaptive value whatsoever during the time of the environmental disruption. An actual example could be the impact on the Earth of a large asteroid from space, such as the Chicxulub impactor that struck at the end of the Cretaceous period of geological time. Within the area of direct blast effects, the adaptive benefits of the morphologies possessed by all animals and plants ceased to exist, as they were vaporized.

The collapse of an adaptive peak results in the death of all the organisms occupying that position within the adaptive landscape. Such an event is termed a *local extinction*. If, however, all the members of a species happen to experience the same lethal environmental disruption, such that none had the good fortune to be in another area of the Earth where the adaptive peak did not collapse entirely, then the entire species goes extinct and a genetic lineage of life ceases to exist. If the

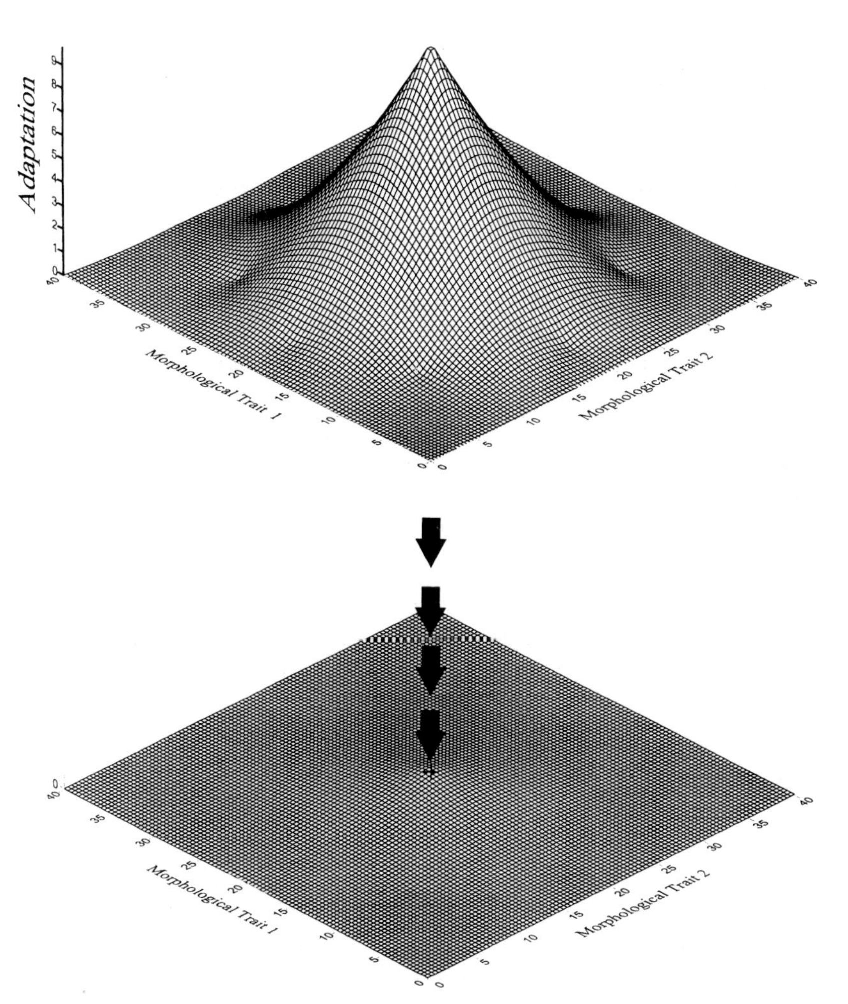

Figure 2.18. Modelling extinction via adaptive peak collapse in an adaptive landscape. In Fig. 2.16 extinction was modelled as the inability of a species to evolve fast enough to keep pace with the adaptive demands of changing environmental conditions. Extinction triggered by major, or catastrophic, environmental disruption can be modelled by collapsing an adaptive peak entirely, rather than having it move across the adaptive landscape. Thus, where there once existed a high altitude peak (top figure) there now exists only a flat plain of zero adaptation (bottom figure). Morphologies that used to work very well (top figure) now have no adaptive value whatsoever (bottom figure) during the time of the environmental disruption.

environmental disturbance is so severe that the entire planet is affected, such as occurred in the asteroid impact at the end of the Cretaceous, then very large numbers of adaptive peaks collapse within the landscape and the result is termed a *mass extinction*.

This last scenario – the extinction of a species – has a drastic but very interesting outcome if that species was the only one possessing the adaptive morphologies associated with the original adaptive peak. When the effects of the lethal environmental disruption have passed, the previous adaptive peak will reappear in the adaptive landscape. Except now the peak is empty: all of the individuals of the species that used to exist in that region of the landscape are gone, and the species is extinct. The possibility is now open for another species to evolve morphologies very similar to the extinct species, and to occupy the vacant adaptive peak. Such a phenomenon is called *convergent evolution*, and we shall examine it in more detail in the next chapter.

3

Modelling evolutionary phenomena in adaptive landscapes

> Simpson (1944) boldly used an adaptive landscape to synthesize genetical and paleontological approaches to evolution ... No visualization before or since 1944 has been so successful in integrating the major issues and themes in phenotypic evolution.
>
> *Arnold, Pfrender and Jones (2001, p. 9)*

Evolution in geological time

In the previous chapter we have seen that the adaptive landscape concept allows us to easily visualize and model the possible effects of natural selection in evolution through simple spatial relationships. As those models became more detailed, we began to encounter larger scale evolutionary phenomena, such as mass extinction and convergent evolution, that may involve thousands of species or operate across millions of years of time.

The first morphologist to use Sewall Wright's concept extensively to model evolutionary phenomena in adaptive landscapes was the palaeontologist George Gaylord Simpson in his classic books *Tempo and Mode in Evolution* (1944) and *The Major Features of Evolution* (1953). Wright's original concept was that of a fitness landscape; movement across that landscape involves changes in genotypic frequencies, small scale changes that are termed *microevolution*. Simpson made the conceptual jump from the fitness landscape of genotypes to the adaptive landscape of phenotypes, or morphologies, and the jump to large-scale evolutionary phenomena that operate on timescales of millions of years, or *macroevolution*. In this chapter we shall examine some of the large scale evolutionary phenomena that have been preserved in the fossil record of life, and to explore the possibility of modelling these patterns of evolution in adaptive landscapes.

Modelling convergent evolution

One of the most striking phenomena that we observe in the fossil record is that of convergent evolution, where a number of different species of animals or plants evolve morphologies that are very similar to one another, even though these species may be only very distantly related and originally started out with ancestral morphologies that were very different from one another. The phenomenon of convergent evolution is one of the most powerful sources of evidence that we have for adaptive evolution, for evolution by the process of natural selection. It is entirely possible, given the immensity of geological time, that two or more species may evolve similar morphologies purely by chance, as is hypothesized in neutralist or random models of evolution. It is also entirely possible that two or more species may evolve similar morphologies, particularly if the species are fairly closely related, simply due to developmental constraint; that is, due to the fact that they have a limited number of possible ways in which they can develop given their particular genetic legacy; sooner or later two species will use the same developmental pathway simply by chance. But it is highly unlikely that large numbers of species of organisms will repeatedly evolve similar morphologies, over and over again in time, simply by chance. Yet that is precisely what we observe in the fossil record: the convergent evolution of form in many different groups of organisms, over and over again in time.

We can easily model, and understand, the phenomenon of convergent evolution using adaptive landscapes. Consider one of the more striking examples of convergent evolution: the similarity in form between an ichthyosaur, a porpoise, a swordfish and a shark. An ichthyosaur is a reptile and a porpoise is a mammal, animals that are very different from one another, yet both look strikingly like a swordfish or shark, streamlined and fusiform (Fig. 3.1). All vertebrate forms of life on land are the descendants of Devonian lobe-finned fish, yet the ichthyosaur is the descendant of a group of land-dwelling reptiles that evolved swimming adaptations and went back into the oceans some 150 million years later, during the Mesozoic. Likewise, the porpoise is the descendant of a group of land-dwelling mammals that also have

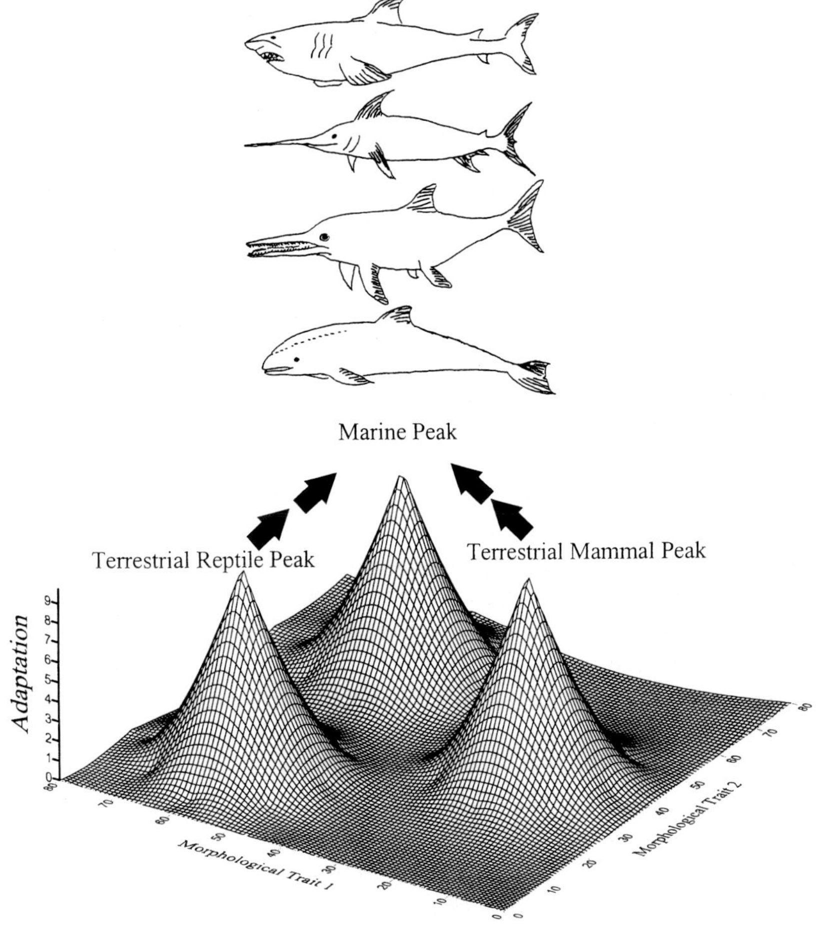

Figure 3.1. Modelling convergent evolution in adaptive landscapes. Although terrestrial mammals and terrestrial reptiles are morphologically very different, occupying distinctly different adaptive peaks within the landscape, they have convergently evolved species that strongly resemble sharks or swordfish. The marine adaptive peak is that of streamlined, fusiform morphologies highly adapted for fast swimming, and it has been heterochronously reached by cartilaginous fishes (sharks, top animal in the figure), bony fishes (swordfish, second down in the figure), marine reptiles (ichthyosaurs, third down in the figure), and marine mammals (porpoises, fourth down in the figure) in different periods of geological time (fish in the Palaeozoic, ichthyosaurs in the Mesozoic, and porpoises in the Cenozoic).
Source: Animal sketches redrawn from Funk and Wagnall (1963).

secondarily returned to the oceans today, and that live a life much like their ancient ancestors, the fish.

Thus, even though the original starting positions for a terrestrial reptile and a terrestrial mammal are very different (Fig. 3.1), the final destination for both the ichthyosaur and porpoise is in the same region of the adaptive landscape (Fig. 3.1), namely on top of the adaptive peak for morphologies that function well in active swimming. That is, morphologies that are streamlined and fusiform, with flippers and fins for powered swimming and steering (even though the ancestors of both the ichthyosaur and the porpoise possessed legs and feet for walking on dry land). It is no accident that both the ichthyosaur and the porpoise look like fish, even though they are not.

The phenomenon of convergent evolution means that there are a limited number of ways of making a living in nature, a limited number of ways of functioning well in any particular environment. We can model this reality in an adaptive landscape by specifying the location of adaptive peaks for particular ways of life: the adaptive peak for an ocean-dwelling active-swimming animal is located in the streamlined fusiform morphological region of the landscape. No matter where you begin your journey on the landscape, if you evolve to be a active-swimming oceanic animal you will wind up in the same region of the landscape; that is, you will converge on the same morphological solution, the same adaptive peak (Fig. 3.1).

The evolutionary convergence in form between the ichthyosaur and the porpoise is an example of *heterochronous* convergence, of convergence on the same morphological solution that nevertheless has taken place at different points in time (during the Mesozoic for the ichthyosaur and at the present time for the porpoise). The phenomenon of heterochronous convergence is extremely common in the fossil record; living organisms repeatedly rediscover the same morphological solutions to the same adaptive conditions, over and over again.

Isochronous convergence, convergence on the same form by two different groups of organisms at the same point in time, also takes place in evolution. Convergence on the same adaptive morphology is the evolutionary consequence of convergence on the same way of life, and two groups of organisms with exactly the same way of life are in danger of intense competition if they live in the same area. Thus, in many cases of isochronous convergence the two groups of evolving organisms are separated in space, rather than time. A striking example is the evolution of sabre-tooth true cats (such as *Homotherium* and

Smilodon) in the old world and North America, and sabre-tooth marsupial 'cats' (such as *Thylacosmilus*) in geographically isolated South America, during the later Cenozoic (Fig. 3.2). The true cats are placental mammals and are very different from marsupial mammals. Yet both placental and marsupial large-game predators evolved sabre-tooth cat-like morphologies that are very similar, though in different regions of the Earth, during the late Cenozoic. Another example of isochronous convergence on morphology and way of life can be seen in flying insectivores: birds and bats look very similar because both have

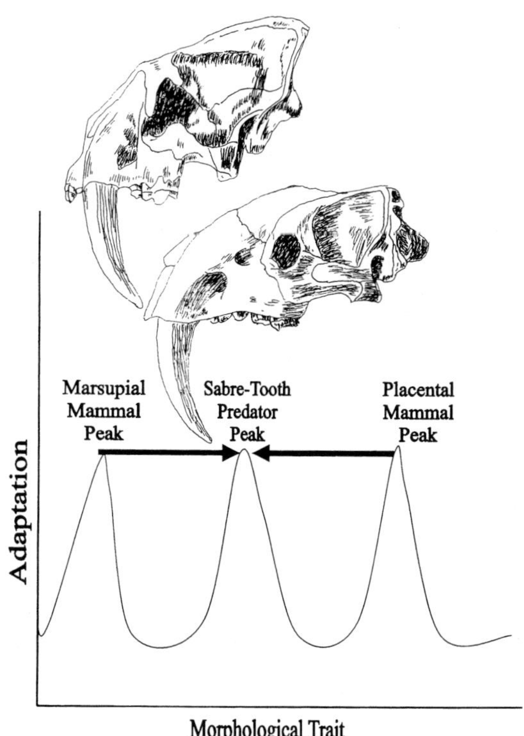

Figure 3.2. Modelling convergent evolution in adaptive landscapes. Placental mammals and marsupial mammals simultaneously evolved predators capable of bring down large prey animals, resulting in the isochronous convergent evolution of sabre-tooth cat-like morphologies (skull of the true cat *Smilodon* at top left, and the marsupial 'cat' *Thylacosimlus* at lower right) during the later Cenozoic, but in different regions of the world (marsupials in South America, and cats in the remainder of the world).
Source: Skull sketches redrawn from Kuhn-Schnyder and Rieber (1986).

evolved wing morphologies for flight, but they are very different kinds of animals, as birds are the descendants of theropod dinosaurs and bats are placental mammals. Although the two groups of animals occupy the same adaptive peak at the same point in time and in the same geographic region, competition is minimized for the two groups by hunting for food during different periods of the day: bats capture insects during the twilight and night, insectivorous birds hunt during the day.

Modelling iterative evolution

Another striking phenomenon observed in the fossil record is that of iterative evolution. Iterative evolution is not as common in the evolution of life as convergent evolution, though at first glance it appears to be very similar. In iterative evolution a group of daughter species with very similar morphologies repeatedly originate, one after another, from an ancestral species that in itself may change very little in geological time. Why is this not simply an example of heterochronous convergence? In a way it is, as each daughter species converges on the same morphology, and the daughter species are separated from one another in time. The important difference from normal convergent evolution is the fact that the heterochronous convergent species all originate from the same ancestral species, thus the phenomenon is more the iteration of a similar morphology from a single source rather than the convergence on a similar morphology from multiple sources.

We can model iterative evolution with a single stable adaptive peak, representing the morphological position of the ancestral species, and a region in the landscape containing an unstable adaptive peak that repeatedly collapses and reappears, representing the iterative morphological position of the daughter species. Iterative evolution was particularly common in Mesozoic ammonites, swimming cephalopods with intricate chambered shells (Bayer and McGhee, 1984, 1985; McGhee, Bayer and Seilacher, 1991). An ancestral, open-ocean ammonite species often persisted for long periods of time during the Mesozoic, as its deep water oceanic environment rarely changed. As sea-level rose and fell during the Mesozoic, however, shallow water habitats on the margins of the continents repeatedly appeared and disappeared, and these habitats were repeatedly invaded by ammonites from the deep sea (Fig. 3.3). The environmental conditions present in shallow-water habitats are very

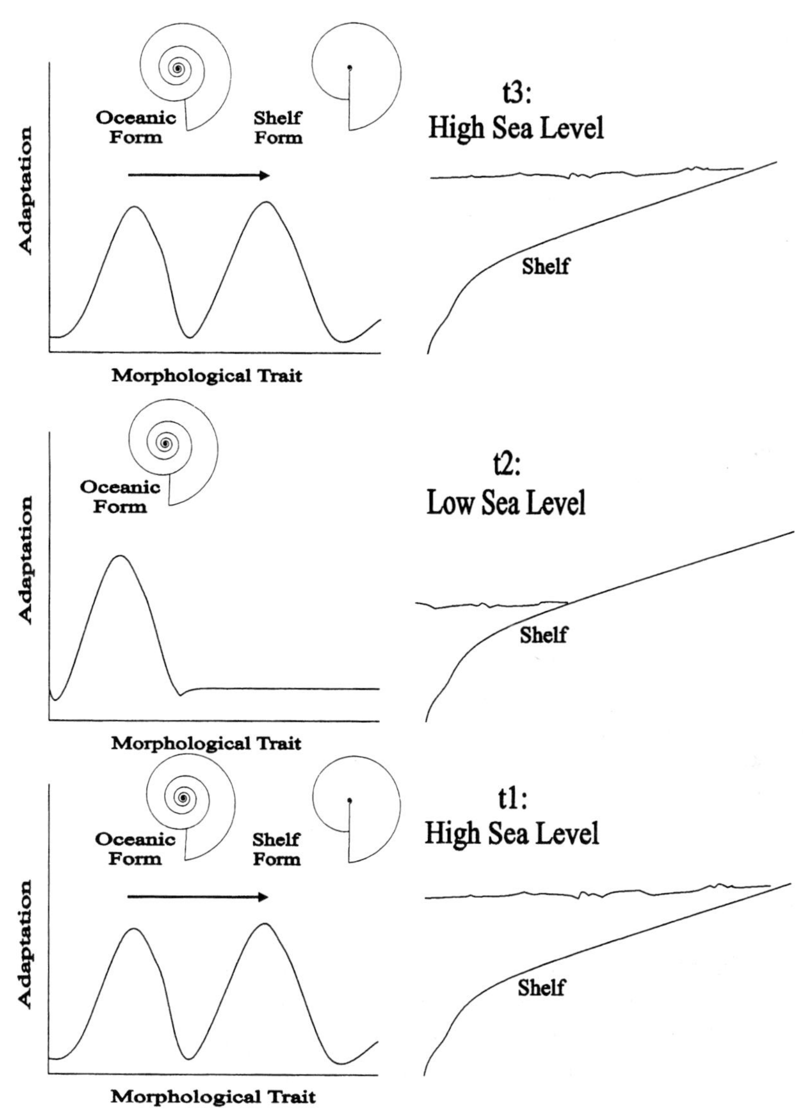

Figure 3.3. Modelling iterative evolution in adaptive landscapes. In the Mesozoic, deep-water oceanic-form ammonites repeatedly evolved shelf-form ammonite species, with very similar morphologies, that inhabited shallow-water habitats during times of high sea level. The iterative evolution of a shelf-form ammonites may be modelled by the appearance of an adaptive peak during a period of high sea level at time = t1, the collapse of the peak (and extinction of the shelf-form ammonite) with the retreat of the seas at time = t2, and the reappearance of a very similar shelf-form ammonite when the adaptive peak reappears in the next cycle of sea-level rise (time = t3).

different from those in the open ocean, however, and the ammonites evolved new shell forms to better function in the new hydrodynamic conditions they found themselves in.

Sooner or later, sea-level would again fall during the Mesozoic, the seas would retreat from the continental margins, and the shallow-water habitats would vanish. In essence, the adaptive peak has collapsed, and the ammonites occupying that peak became extinct (Fig. 3.3). In the next cycle of sea-level rise, the shallow-water habitats reappeared with the same hydrodynamic conditions as in the previous period of high sea level; that is, the same adaptive peak reappeared on the landscape (Fig. 3.3). And once again, these habitats were invaded by deep water ammonites that evolved shell forms very similar to the previous, but now extinct, ammonites.

Modelling speciation by cladogenesis

For many years a noisy debate raged concerning the true pattern of cladogenesis (the phenomenon of species lineage splitting) as observed in the fossil record (it still continues, but is quieter these days). The debate concerned the 'phyletic gradualism' model of speciation versus the 'punctuated equilibrium' model (Fig. 3.4). The essence of the debate concerns the question of how much morphological change occurs *within a species lineage* (called anagensis) during the existence of the species versus how much morphological change occurs *between species lineages* (called cladogenesis) in the speciation event. In the phyletic gradualism model, morphological evolution consists of roughly equal amounts of cladogenetic change and anagenetic change (Fig. 3.4). In the punctuated equilibrium model, morphological evolution consists almost entirely cladogenetic change, with very little anagenetic change (Fig. 3.4). Species lineages, once established, are modelled as exhibiting no morphological evolution over long periods of time, a phenomenon known as *morphological stasis*.

The geometry of the two patterns of speciation appear to be radically different (Fig. 3.4), thus at first glance it would seem easy to prove one or the other. But not so. In general, many biologists (students of present life) tended to favour the phyletic gradualism model and maintained that Darwin himself had been a gradualist (complete with quotes from the authoritative book, *On the Origin of Species*). They also maintained that the fossil record was very fragmentary (as indeed Darwin did

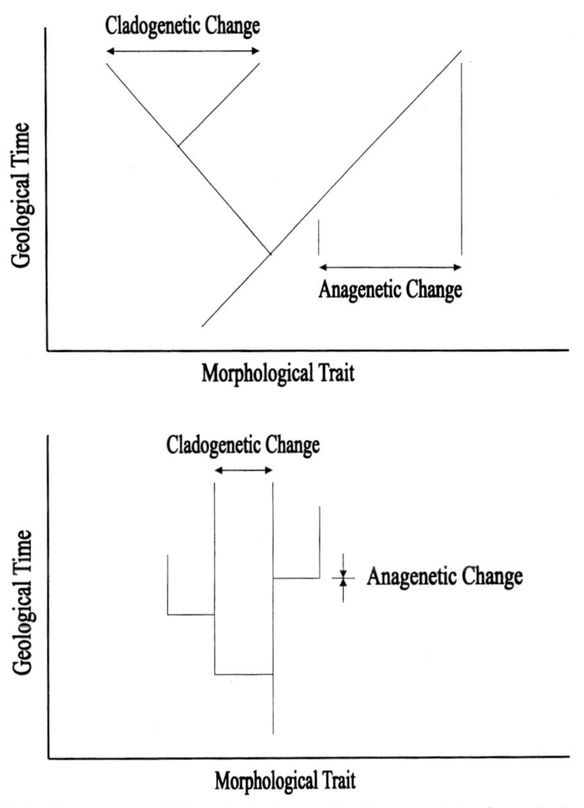

Figure 3.4. Geometry of the phyletic gradualism model of speciation (top) and of the punctuated equilibrium model of speciation (bottom). In the phyletic gradualism model morphological evolution consists of roughly equal amounts of cladogenetic change between lineages and anagenetic change within lineages, whereas in the punctuated equilibrium model morphological evolution consists almost entirely of cladogenetic change.

himself), and that if you simply looked hard enough for the best preserved fossil record of speciation, it would turn out to match the phyletic gradualism style of geometry (Fig. 3.4). On the other hand, many palaeontologists (students of past life) tended to favour the punctuated equilibrium model. They maintained that the fossil record in general was pretty good, and that if you examined many different examples of speciation in the fossil record, they generally displayed the geometry in the punctuated equilibrium model (Fig. 3.4).

The palaeontologist George Gaylord Simpson (1944, 1953) first explicitly modelled morphological evolution and speciation in adaptive landscapes over a half a century ago. It is instructive to examine his ideas in an adaptive landscape context, relative to the debate concerning the phyletic gradualism versus punctuated equilibrium models of evolution. In Figure 3.5 is his model of equid (horse) evolution in the Cenozoic. The equids started out as small browsing animals in the Eocene (the Hyracotheriinae), and later evolved into much larger animals (the Anchitheriinae), some of which made the transition from a browsing mode of feeding to a grazing mode of feeding (the Equinae).

Note that Simpson modelled the evolution of the Anchitheriinae from the Hyracotheriinae as a process produced by shifting adaptive peaks (as in Fig. 2.15 of the last chapter) without lineage splitting; that is, as an anagenetic process. He then modelled the cladogenetic evolution of the Equinae as a process of disruptive selection (as in Figs. 2.7–2.10), with the Equinae splitting away from the Anchitheriinae as they became more and more adapted to a grazing mode of feeding by climbing the grazing adaptive peak. Thus Simpson clearly considered the morphological evolution of the equids to entail both anagenetic and cladogenetic change.

Consider a second example: Simpson's model of penguin evolution in an adaptive landscape context (Fig. 3.6). Simpson now models both of the the aerial-flight adaptive peaks, aquatic and terrestrial, as unchanging and stationary with time (unlike the shifting of the browsing adaptive peak modelled in Fig. 3.5) and the lineages occupying those peaks as morphologically static. Although he does model the two other flight styles as shifting adaptive peaks, particularly the submarine flight peak (Fig. 3.6), most of penguin morphological evolution is modelled as a process of jumping from one adaptive peak to another. That is, he modelled morphological evolution within the penguins as a punctuated, cladogenetic process (Fig. 3.6).

Simpson's two models of speciation are summarized in Figures 3.7 and 3.8. Speciation as a process of disruptive selection on two adaptive peaks is modelled in Figure 3.7; such a process should produce a pattern of morphological evolution that consists of both cladogenetic change between the two diverging lineages and anagenetic change within the two lineages as they climb their respective adaptive peaks. Thus the disruptive selection model clearly supports the phyletic gradualism model of evolution.

Speciation as a process of adaptive peak jumping is modelled in Figure 3.8. Such jumps must of necessity be very quick; once firmly established on a new adaptive peak the new species should experience

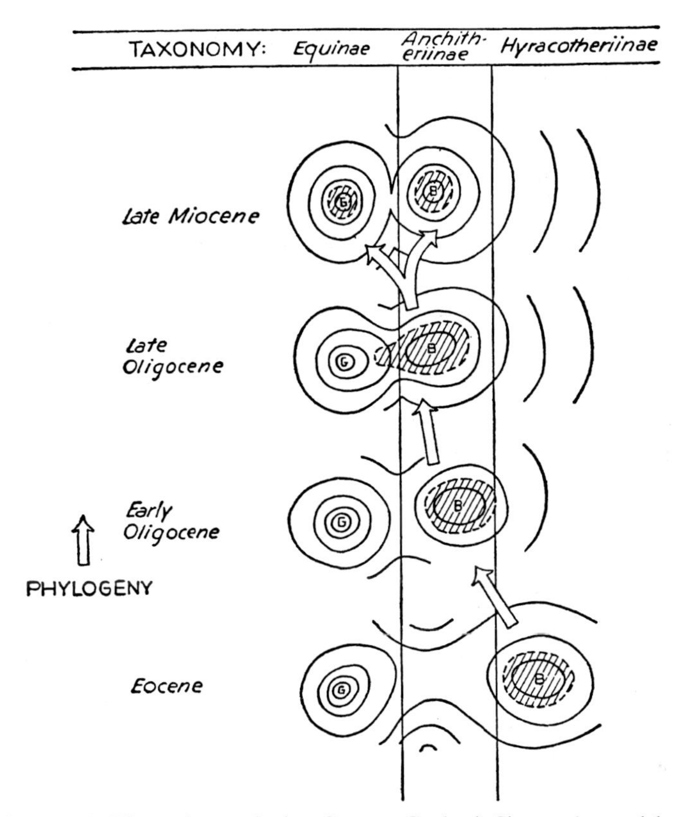

Figure 3.5. The palaeontologist George Gaylord Simpson's model of morphological evolution in the *Equidae* (the horse family) in the Cenozoic. Members of the equid subfamily *Hyracotheriinae* are shown to occupy an adaptive peak (shaded oval in the right bottom corner of the figure), labelled 'B' for teeth morphologies adapted for browsing, during the Eocene. In the left bottom corner of the figure is illustrated an empty adaptive peak, labelled 'G' for teeth morphologies adapted for grazing. In the Oligocene the subfamily *Anchitheriinae* are shown to evolve anagenetically from the earlier *Hyracotheriinae*, and the browsing adaptive peak is shown to be moving closer to the empty grazing adaptive peak, due to modifications in the teeth morphologies of the *Anchitheriinae* due to their evolution of much larger body masses and sizes than the *Hyracotheriinae*. In the Miocene the original *Anchitheriinae* lineage has split in two, with one subgroup evolving teeth morphologies that enable it to occupy the grazing adaptive peak and resulting in the cladogenetic evolution of the subfamily *Equinae* from the *Achitheriinae*.
Source: From *The Major Features of Evolution*, by G. G. Simpson. Copyright © 1953 by Columbia University Press and reprinted with the permission of the publisher.

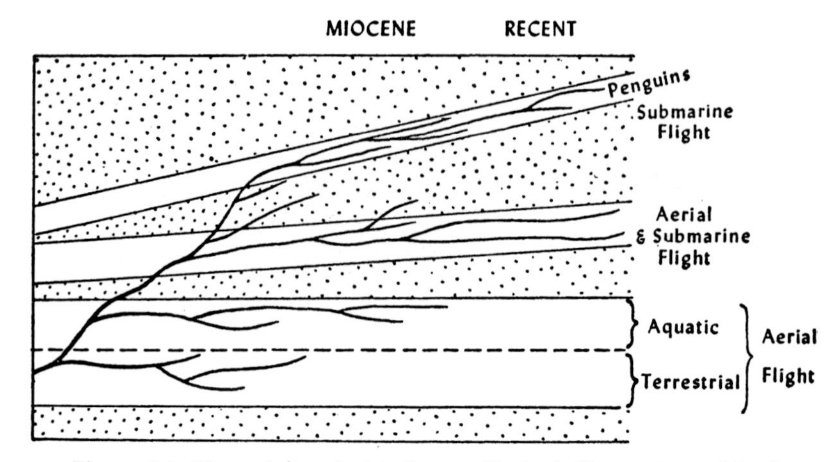

Figure 3.6. The palaeontologist George Gaylord Simpson's model of morphological evolution in the penguins. Rather than a topographic map portrayal of the adaptive landscape (as in Fig. 3.5), the positions of the adaptive peaks for differing flight styles are shown as unshaded zones, or corridors, within a shaded landscape field of less-adaptive morphologies. The horizontal dimension is time; note that Simpson depicts the aquatic and terrestrial adaptive-peak zones as unchanging through time; these peaks are modelled as stationary. Note further that the two other flight-style adaptive-peak zones are drawn at an angle to the horizontal time dimension, thus these peaks are modelled as shifting with time. The evolution of penguin morphologies adapted to the spectrum of flight types, from aerial to submarine, is modelled as a series of cladogenetic events as the ancestral population splits and jumps from one adaptive-peak zone to another.
Source: From *The Major Features of Evolution*, by G. G. Simpson. Copyright © 1953 by Columbia University Press and reprinted with the permission of the publisher.

very little further morphological change due to stabilizing selection (as modelled in Figs. 2.5–2.6 of the last chapter). Thus the peak-jumping model clearly supports the punctuated equilibrium model of evolution, with the majority of morphological change occurring in the cladogenetic origination of the new species, which then exhibits morphological stasis (i.e. no anagenetic morphological change) from then on.

Can we now conclude that evolution via the process of disruptive selection will *always* produce a phyletic gradualism pattern, while evolution via the process of peak jumping will *always* produce a punctuated equilibrium pattern? Not really. The key flaw in that

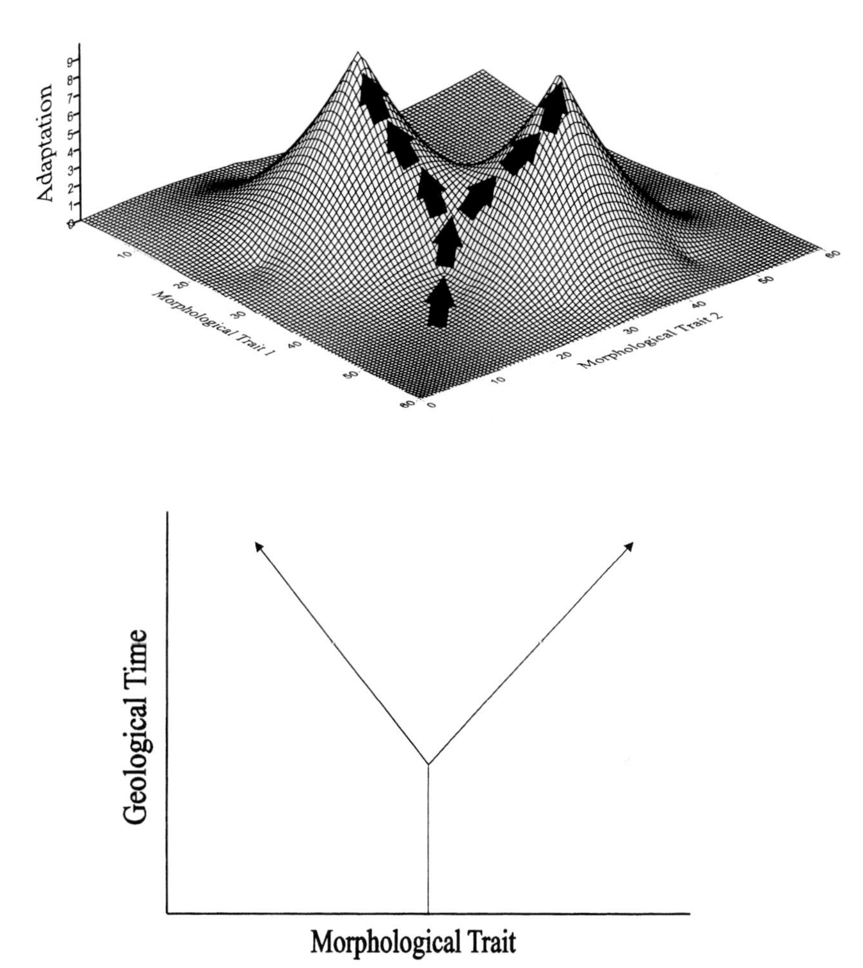

Figure 3.7. Modelling cladogenesis as a product of disruptive selection. Illustrated in the top figure is an ancestral species population that splits into two descendant species that climb two different adaptive peaks, as modelled in Figure 2.10. Illustrated in the bottom figure is a possible model of the geometry of evolution produced by disruptive selection, where morphological change consists of roughly equal amounts of cladogenetic change between the two species lineages and anagenetic change within the two lineages.

conclusion is the apparent slow divergence of the two species as they climb the two adaptive peaks in Figure 3.7. If that divergence is not slow at all, but in actual life takes place in a few hundred to thousand years, then in the fossil record the speciation event will appear to be

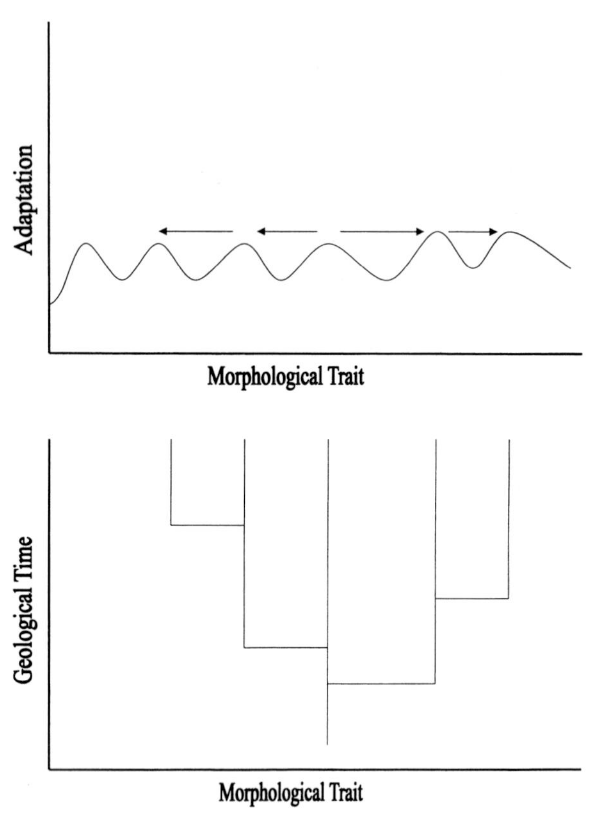

Figure 3.8. Modelling cladogenesis as a product of peak jumping. Illustrated in the top figure is an ancestral species population that produces two descendant species by the process of peak jumping, which in turn produce two more descendant species by the same process, as modelled in Figure 2.14. Illustrated in the bottom figure is a possible model of the geometry of evolution produced by peak jumping, where morphological change consists almost entirely of cladogenetic change.

virtually instantaneous and indistinguishable from the pattern produced by the peak-jumping model (Fig. 3.8). Indeed, in the actual speciation debate many biologists and palaeontologists came to realize that they were misunderstanding each other's concept of the rapidity of the evolutionary process. Once the biologists came to understand that the palaeontologists were not calling for instantaneous speciation in ecological time (as in one day or week!) but in geological time (as in several hundred to a thousand years), this entire part of the debate collapsed.

The debate still continues, but now in a direction of more significance to modelling evolution with adaptive landscapes. The phyletic gradualism model is more compatible with an adaptive landscape that is constantly in motion, such that a significant part of evolutionary change occurs not only in cladogenetic (lineage splitting) events, but in anagenetic morphological change as species evolve simply to follow the adaptive peaks that they occupy as those peaks move about the landscape.

On the other hand, the punctuated equilibrium model is more compatible with an adaptive landscape where the peaks are stationary for long periods of time, and then move in an abrupt or quantal fashion, or abruptly collapse and reappear. Such a landscape is predicted by the long periods of morphological stasis that occurs within each species lineage in the punctuated equilibrium model (Fig. 3.4), where virtually no anagenetic change occurs at all. Thus, rather than the pattern of cladogenetic change, it is the debate over the pattern of anagenetic change in geological time that is now of interest to modellers of the process of evolution in adaptive landscapes.

Modelling hyperdimensionality in adaptive landscapes

Real life is very complex and it is clear that an actual fitness, or adaptive, landscape must of necessity be a hyperdimensional space. The geneticist Theodosius Dobzhansky, who greatly admired Sewall Wright's concept, nevertheless realized that an actual fitness landscape must possess a staggering number of dimensions:

> Suppose there are only 1000 kinds of genes in the world, each gene existing in 10 different variants or alleles. Both figures are patent underestimates. Even so, the number of gametes with different combinations of genes potentially possible with these alleles would be 10^{1000}. This is fantastic, since the number of subatomic particles in the universe is estimated as a mere 10^{78}...
>
> *(Dobzhansky, 1970, p. 25)*

Dobzhansky's estimate of the potential number of possible genetic combinations in nature is a poignant example of the complexity of biology as opposed to physics. We now know, from the results of the Human Genome Project, that the human genome itself contains some 20000 to 25000 genes, a number much greater than the 1000 that Dobzhansky offered as a tentative estimate of the number of genes in the entire world.

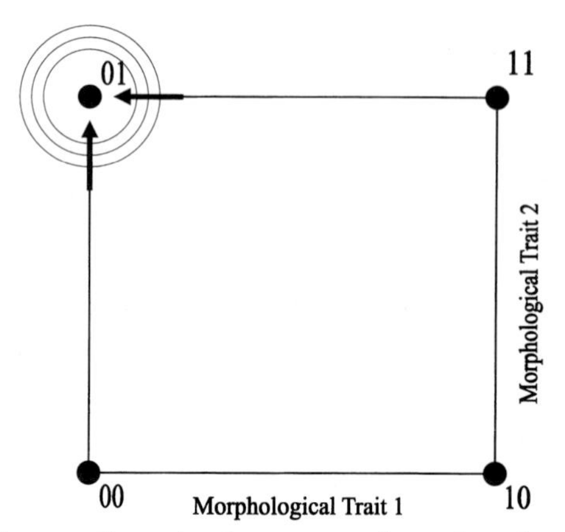

Figure 3.9. A two dimensional representation of a three-dimensional adaptive landscape. The two dimensions of the plane of the page are morphological dimensions, and the third dimension of adaptation is given by the adaptive contours, which measure vertical height above the plane of the page. The two vectors represent the action of natural selection, in which evolving populations climb the adaptive peak.

Thus far we have modelled adaptive landscapes in three dimensions, two of morphology and one of adaptation. Let us now consider modelling higher dimensional adaptive landscapes. Figure 3.9 shows the contour-map model of an adaptive landscape that we have used thus far in this book. Let us now give positional coordinates to the morphological dimensions, where morphological traits one and two can vary from a value of zero to a value of one. The four corners of the landscape thus have the coordinates 00, 10, 01 and 11 (Fig. 3.9). In the lower left corner, 00, neither morphological trait one nor two is present. In the lower right corner, 10, the first dimension, morphological trait one, is present in its maximum value in hypothetical organisms that might occur in this region of the landscape, but morphological trait two is absent in these same organisms. In the upper left corner, 01, the opposite is true. Here the second dimension, morphological trait two, is present in its maximum value in hypothetical organisms that might occur in this region of the landscape, whereas morphological trait one is absent in these same organisms. Last, in the upper right corner, 11, hypothetical organisms in this region of the landscape possess both morphological traits developed in their maximum value. Thus in our coordinate notation, the first digit represents the coordinate value of the first morphological dimension, and the second digit represents the value of the second morphological dimension.

The third dimension in Figure 3.9 is adaptation, represented by the adaptive contours that measure vertical height above the two-dimensional plane of the morphological dimensions. In Figure 3.9 an adaptive peak is shown to occur in the upper left corner of the landscape, centred on the morphological coordinates 01. The two vectors show the predicted evolutionary pathway of morphological evolution in the two morphological dimensions, from the expectations of the theory of natural selection.

Let us now add a third morphological dimension to produce a four-dimensional adaptive landscape. Our previous landscape, Figure 3.9, now is the front face of the cube shown in Figure 3.10. Note now that the coordinates of its corners are no longer 00, 10, 01, and 11, but rather 000, 100, 010, and 110 (Fig. 3.10). The third digit, zero in all of the corners in the front face of the cube, is the value of the third dimension, morphological trait three. If we move into that third morphological dimension we see that the morphological coordinates of the rear face of the cube are 001, where hypothetical organisms in this region of the four-dimensional landscape possess only morphological

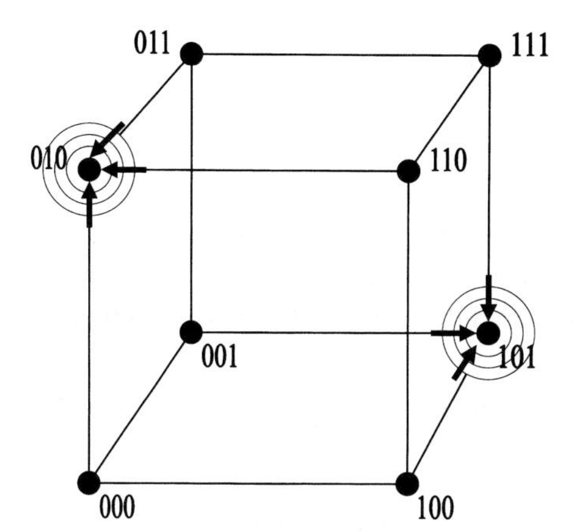

Figure 3.10. A three dimensional representation of a four-dimensional adaptive landscape. The three dimensions of the cube are morphological dimensions, and the fourth dimension of adaptation is given by the adaptive contours, which now are concentric spheres with surfaces of equal adaptive value. The three vectors represent the action of natural selection, in which evolving populations climb to the centre of the concentric adaptive spheres.

trait three, 101, the region where organisms have both morphological traits one and three, 011, the region where organisms have both morphological traits two and three, and 111, the region where organisms possess all three morphological traits.

The fourth dimension of the landscape is adaptation, represented by the adaptive contours. Note now that the contours represent the surfaces of concentric spheres, nested one within the other, where the surface of any one sphere represents morphological coordinates that have equal adaptive value. The adaptive peak centred on coordinates 01 in the three-dimensional landscape in Figure 3.9 now is shown centred on the coordinates 010 in the four-dimensional adaptive landscape (Fig. 3.10). We now discover that another adaptive peak exists, centred on the morphological coordinates 101, that we could not see when we were confined to seeing only in three dimensions (Fig. 3.9). The three vectors drawn on both adaptive peaks show the predicted evolutionary pathway of morphological evolution in the three morphological dimensions, in which evolving populations climb in the fourth dimension to the centre of the concentric adaptive spheres.

Let us now add a fourth morphological dimension to produce a five-dimensional adaptive landscape. Our previous landscape, Figure 3.10, now is the outer cube of the *hypercube* shown in Figure 3.11. Note now that the coordinates of its corners are now longer 000, 100, 010, and 110 (Fig. 3.10), but rather 0000, 1000, 0100, and 1110. The fourth digit, zero in all of the corners in the outer cube of the hypercube, is the value of the fourth dimension, morphological trait four. If we move into that fourth morphological dimension we see that the morphological coordinates of the front face of the inner cube of the hypercube are 0001, where hypothetical organisms in this region of the five-dimensional landscape possess only morphological trait four, 1001, the region where organisms have both morphological traits one and four, 0101, the region where organisms have both morphological traits two and four, and 1101, the region where organisms possess morphological traits one, two and four. The morphological coordinates of the rear face of the inner cube of the hypercube are 0011, where hypothetical organisms in this region of the landscape possess morphological traits three and four, 1011, the region where organisms have morphological traits one, three and four, 0111, the region where organisms have morphological traits two, three and four, and finally 1111, the region of the

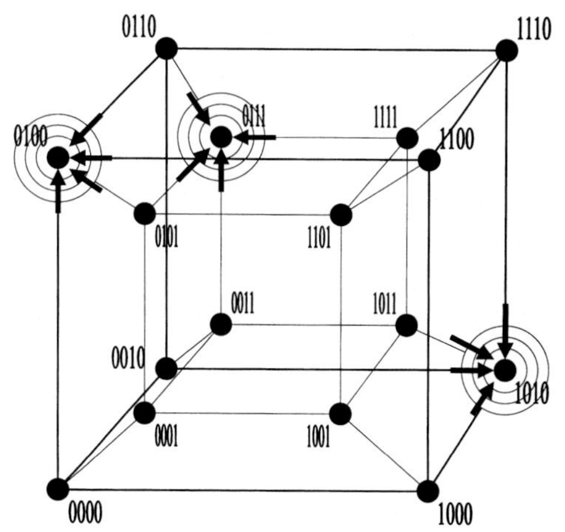

Figure 3.11. A four dimensional representation of a five-dimensional adaptive landscape. The four dimensions of the hypercube are morphological dimensions, and the fifth dimension of adaptation is given by the adaptive contours, which now are concentric spheres with surfaces of equal adaptive value. The four vectors represent the action of natural selection, in which evolving populations climb to the centre of the concentric adaptive spheres.

five-dimensional landscape where organisms possess all four morphological traits.

The fifth dimension of the landscape is adaptation, represented by the adaptive contours. Note that the two adaptive peaks centred on coordinates 010 and 101 in the four-dimensional landscape (Fig. 3.10) now are shown centred on the coordinates 0100 and 1010 in the five-dimensional adaptive landscape (Fig. 3.11). We now discover that yet another adaptive peak exists, centred on the morphological coordinates 0111 (Fig. 3.11), that we could not see when we were confined to seeing only in four dimensions (Fig. 3.10). The four vectors drawn on all three adaptive peaks show the predicted evolutionary pathway of morphological evolution in the four morphological dimensions, in which evolving populations climb in the fifth dimension to the centre of the concentric adaptive spheres.

This is as far as we can go with simple visual representations of hyperdimensional spaces. Mathematically, however, there is no reason to stop at five dimensions. Although we cannot visually portray a six-dimensional adaptive landscape, as we did with a five-dimensional space in Figure 3.11, we still can explore its properties mathematically. One interesting mathematical technique is to consider the morphological coordinates given in Figures 3.9 to 3.11 as *only to exist* in a zero or one condition, rather than a morphological spectrum of values from a minimum of zero to a maximum of one. If we switch to a binary, base two, system of mathematics we may then use the techniques of Boolean algebra to explore the mathematical properties of evolution within hyperdimensional adaptive landscapes. Or we can assign hypothetical fitness, or adaptive values, to the Boolean coordinates of the hyperspace, and computer simulate expected evolutionary pathways within the hyperdimensional adaptive landscape. Many of the interesting evolutionary insights of the theoretician Kauffman (1993, 1995), discussed in Chapter 2, were obtained in this fashion.

The question of hyperdimensionality has played an interesting role in evolving concepts of the fitness landscape of genotypes, Sewall Wright's original concept, as discussed in Chapter 1. Wright's evolutionary rival, R. A. Fisher, argued that an inverse relationship existed between the number of adaptive peaks one could expect to find in an adaptive landscape and the number of genotypic dimensions of that landscape (Fisher, 1941). That is, the greater the number of genetic combinational possibilities that one considered, the fewer the number of actually fit combinations one could expect to exist simply due to the sheer number of potential genetic combinations that would be lethal. Thus Fisher argued that the multi-peaked fitness landscape of Wright (1932) was unrealistic and simply the result of the low dimensionality (i.e. two dimensions) of the hypothetical landscape – real fitness landscapes would have very few actual peaks (see discussions in Ridley, 1996 and Futuyama, 1998).

More recently, the hyperdimensionality question has once again arisen, but in a different formulation (Gavrilets, 1997, 1999, 2003; Gavrilets and Gravner, 1997). Gavrilets (1997, p. 307) has argued that the fitness landscape metaphor 'with its emphasis on adaptive peaks and valleys, is to a large degree a reflection of our three-dimensional experience' and argues for 'a new unifying framework that provides a plausible multidimensional alternative to the convential view of rugged adaptive landscapes' (Gavrilets, 2003, p. 135). Using a Boolean-style mathematical model in which the fitness of a genotype is either zero (lethal) or one (fit), Gavrilets and Gravner (1997, p. 51) argue that the resulting fitness landscape is a multimensional 'flat surface with many holes', which they term a *holey fitness landscape*. The 'holes' in the multidimensional space represent the intersection of many planes of genotypic dimensions with those regions of the hyperspace where fit genotypes exist. Rather than modelling evolution as a process of climbing fitness landscapes, in the holey fitness landscape model evolution proceeds by jumping from hole to hole in the landscape through 'extra-dimensional bypasses' (Gavrilets, 1997, p. 311).

This book is concerned with adaptive landscapes and morphology, not with fitness landscapes and gene interactions, so I shall not pursue the dimensionality debate concerning fitness landscapes further than the references mentioned above. However, we shall return to a somewhat similar concept of abrupt, quantal jumps of evolution in potentially discontinuous adaptive landscape surfaces when we consider the questions of morphogenesis and developmental constraint in Chapter 8.

Are adaptive landscapes of heuristic value only?

Although the adaptive landscape concept has been termed a 'standard imagination prosthesis for evolutionary theorists' (Dennett, 1996, p. 190), the many uses of the concept have been mostly that, imaginary or conceptual. As mentioned at the beginning of this chapter, the first morphologist to extensively use the adaptive landscape to model evolution in geological time was George Gaylord Simpson, yet all the examples given in his books *Tempo and Mode in Evolution* (1944) and *The Major Features of Evolution* (1953) are conceptual models. All of the examples of evolutionary processes in adaptive landscapes given in the first three chapters of this book have been conceptual models. And if you survey the biological literature today you will find that the vast majority of mentions of the adaptive landscape concept involve conceptual models, ranging in complexity from sophisticated computer simulations to entertaining graphic cartoons.

We have encountered the concept of rugged adaptive landscapes in Chapter 2. In Figure 3.12 a rugged adaptive landscape for ancient trilobites (extinct marine arthropods) is given from Solé and Goodwin (2000). The portrayed landscape does indeed look rugged, and various peaks are linked to trilobites by arrows. Yet in the text of their work Solé and Goodwin (2000, p. 257) point out that the figure does not have

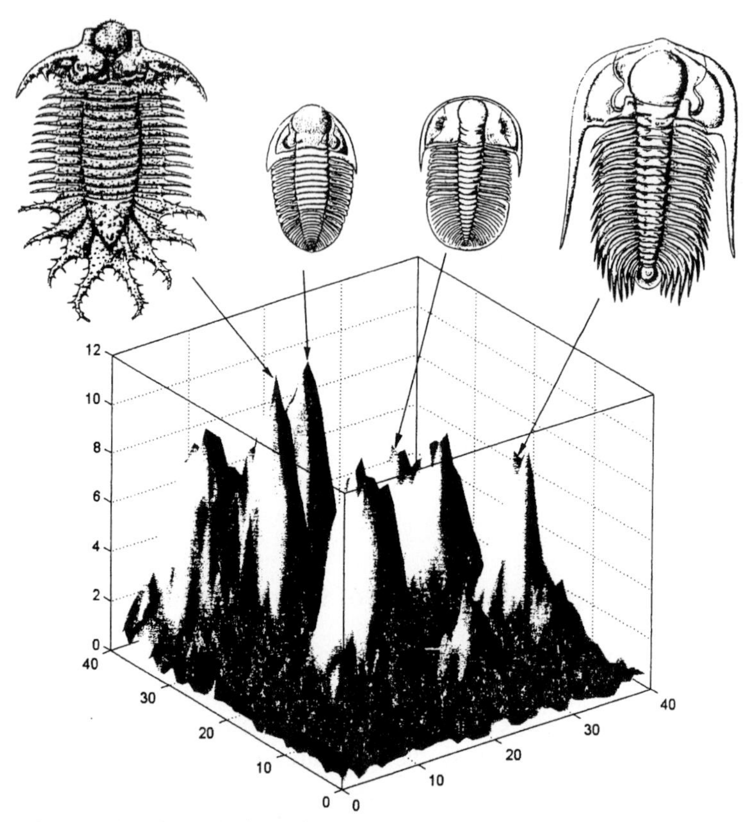

Figure 3.12. The evolution of ancient trilobites (extinct Palaeozoic marine arthropods) in a rugged adaptive landscape. Although the three dimensions of the landscape are labelled with numbers, and the trilobites shown are real organisms, the diagram itself is entirely a conceptual model.
Source: From *Signs of Life: How Complexity Pervades Biology*, by R. Solé and B. Goodwin. Copyright © 2000 by R. Solé and B. Goodwin and reprinted with the permission of R. Solé and B. Goodwin.

anything to do with real trilobites; that is, it is a conceptual model. Another example is given in Figure 3.13, which illustrates an adaptive landscape of eye morphologies from Dawkins (1996), created by his colleague Michael Land. On the left we see mountains representing highly-adapted types of compound eyes, and on the right we see mountains representing highly-adapted variants of camera-type eyes. In the middle ground we see rolling hills represent less highly-adapted types of eyes, and in the foreground we see barely perceptible changes

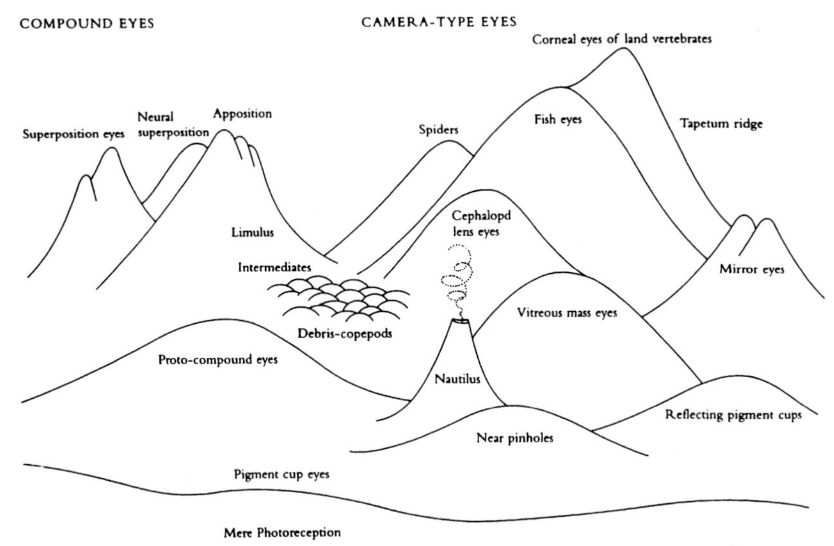

Figure 3.13. The biologist Michael Land's adaptive landscape of eye morphologies. On the left are peaks representing highly-adapted types of compound eyes, and on the right are peaks representing highly-adapted variants of camera-type eyes. In the middle ground are lower peaks represently less highly-adapted types of eyes, and in the foreground are barely perceptible adaptive slopes in the landscape that represent the simplest types of 'eyes' possible, such as mere photoreceptors. Although all of the eye types listed are present in real organisms, the diagram itself is entirely a conceptual model.
Source: From *Climbing Mount Improbable*, by R. Dawkins. Copyright © 1996 by R. Dawkins and reprinted with the permission of R. Dawkins.

in elevation in the landscape that represent the simplest types of 'eyes' possible, such as mere photoreceptors (I have never discovered why Land portrayed the camera-type eyes of the last living ectocochleate cephalopod, *Nautilus*, as a volcano that has blown its top!). This intriguing graphic presents in a single, easily-comprehensible view, the complexity of eye evolution over the past six hundred million years. Yet it is also simply a conceptual model.

One last example of a conceptual model of evolution using adaptive landscapes is given in Figure 3.14, from Strathmann (1978), who has argued that many adaptive-types of organisms may have progressively vanished through the passage of geological time due to the combined effects of occasional peak collapse and of evolutionary specialization.

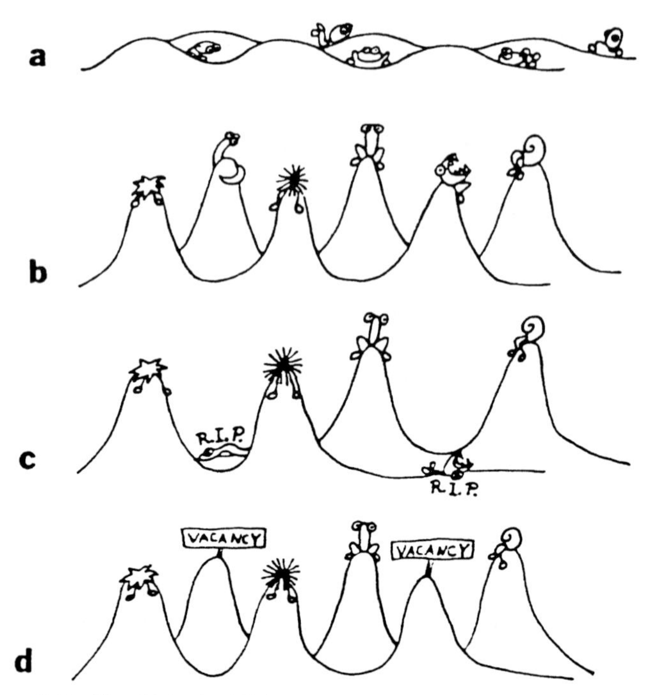

Figure 3.14. The biologist Richard Strathman's conceptual model of the evolution of specialization, and of extinction, in an adaptive landscape. The vertical dimension is time, with the oldest landscape at the top and the youngest at the bottom. Generalist organisms on low peaks in landscape 'a' evolve into specialist organisms on high peaks in landscape 'b'. Two of the peaks collapse in landscape 'c', resulting in the extinction of the organisms that inhabited them. Once the environmental disruption that triggered the peak collapses has passed, the adaptive peaks reappear in landscape 'd' but are now empty. See text for discussion.
Source: From Strathmann (1978). Copyright © 1978 by *Evolution* and reprinted with the permission of the publisher.

In Strathmann's model, organisms evolve from generalists with broad adaptations (animal sketches shown in the top landscape, at time level 'a' in Fig. 3.14) to specialists with narrow adaptations (animal sketches shown on the tops of high peaks, at time level 'b'). Note that Strathmann has not modelled the different organisms climbing pre-existing peaks, but rather has the organisms themselves steepening the slopes of the peaks as they become better and better adapted to their respective environments and ecological setting.

The generalists, located at low altitude positions in the landscape, can move about the landscape with little difficulty as the adaptive valleys are shallow from their relative perspective. Specialists, however, are located in high altitude positions at the very top of adaptive peaks and cannot budge from their high-fitness positions as the adaptive valleys surrounding them are very deep. Under the expectations of the theory of natural selection, Strathmann (1978) argued that most organisms will evolve with time from being generalists to being specialists. Late in the evolutionary scenario most all organisms will be sitting on the tops of high altitude peaks, surrounded by deep valleys (but only if the peaks *remain stable in time*, as we saw in Chapter 2).

Now let us introduce an environmental catastrophe (such as an asteroid falling out of the sky) that results in the temporary destruction of two of the adaptive peaks, which leads to the extinction of the organisms inhabiting those peaks as the fitness of their areas of the landscape has dropped to zero (Fig. 3.14, time level 'c'). After the environment recovers and returns to the precatastrophe state, the two adaptive peaks reappear but are not occupied, as their previous inhabitors have gone (Fig. 3.14, time level 'd').

Now we enter the interesting conclusion of Strathmann's scenario. We have two vacant adaptive peaks on the landscape, two really nice pieces of real estate up for sale (Fig. 3.14, time level 'd'). Surely one or two of the neighbouring organisms will evolve adaptations which will allow them to move onto these vacant adaptive peaks, yes? Strathmann (1978) argued that this may not be possible at this late and specialized stage in the evolution of the organisms. The adaptive valleys which separate the existent organisms from the vacant peaks are simply too deep. The organisms have simply become too specialized to move very far on the adaptive landscape. That is, the conclusion of Strathmann's (1978) argument is that more adaptive-types of life may have existed in the earlier phases of evolution on the Earth than those which exist at present. Rather than steadily increasing the numbers of adaptive-types of life, evolution may have led to the progressive loss of adaptive-types through time. Although that is a model prediction that can be tested by examining the actual numbers of adaptive-types of life seen in the fossil record in geological time, the model producing the prediction (Fig. 3.14) is still that, a conceptual model.

The actual utilization of an adaptive landscape in the actual analysis of genotypes or morphologies is hampered by the complexity of biological form, and for that reason some critics have stated that the adaptive landscape concept is of heuristic value only. That is, it is fine for creating conceptual models, but you cannot actually use the concept in analysing the evolution of actual animals or plants. We shall see in the next chapter, however, that the adaptive landscape concept indeed can be put into practice through the usage of *theoretical morphospaces*.

4

The concept of the theoretical morphospace

> The study of evolutionary constraint requires a metric — a map for visualizing the occupied and unoccupied evolutionary pathways that are theoretically possible. Two such maps are the often cited 'adaptive landscape' of genetic frequencies (from Sewall Wright) and David M. Raup's 'morphospace' of coiled shells.
>
> *Schindel (1990, p. 270)*

What is a theoretical morphospace?

Imagine a room whose floor is covered with beautiful glass models of sea shells. The glass sea-shell models are carefully arranged in a pattern of parallel rows on the floor, such that as you walk down one row after another, you can see the glass models change their geometries progressively from one type of sea shell to another. At the end of one row you might find a glass model that looks very much like a snail shell, but at the end of that same row the spire of the models has become so low that the glass model now looks more like a clam shell. You are walking in a theoretical morphospace (Fig. 4.1).

The concept of the theoretical morphospace was first proposed by the palaeontologist David M. Raup in 1966; it is an extension of the adaptive landscape concept. Theoretical morphospaces may be defined most explicitly, if a bit tersely, as 'n-dimensional geometric hyperspaces produced by systematically varying the parameter values of a geometric model of form' (McGhee, 1991, p. 87). The main difference between the adaptive landscape and the theoretical morphospace lies in their dimensions. The dimensions of the adaptive landscape are morphological

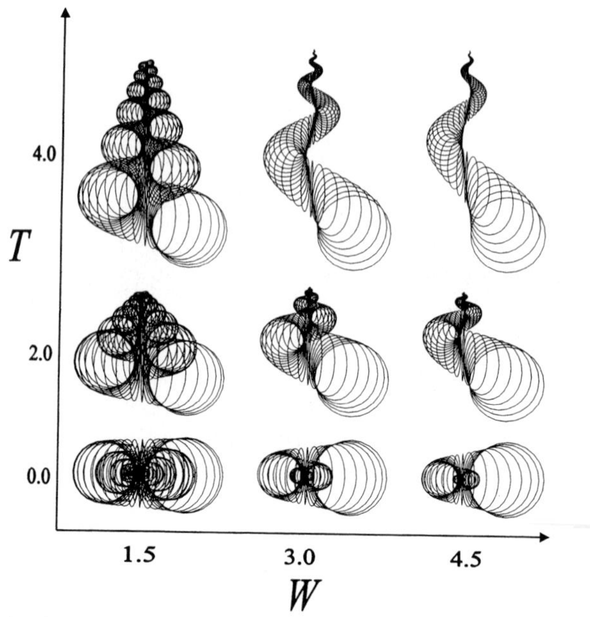

Figure 4.1. A theoretical morphospace of hypothetical mollusc shells. The two morphological trait dimensions are *W*, the whorl expansion rate of the shell, and *T*, the translation rate of the shell.
Source: These computer simulations were produced using a modified version of the source code that may be found in Swan (1999), rewritten in Visual Basic 6.0.

traits and degree of adaptation (Fig. 4.2). The degree of adaptation, or fitness, is a fundamental feature of the adaptive landscape concept.

In contrast, the dimensions of a theoretical morphospace are geometric or mathematical models of form and the frequency of occurrence of those hypothetical model forms in nature (Fig. 4.2). The concept of adaptation does not enter into the construction of a theoretical morphospace, whereas the concept of adaptation is a fundamental feature of an adaptive landscape.

Within an adaptive landscape, different locations not only represent differing morphological or genetic combinations, they represent different degrees of adaptation or fitness. As the dimensions of a theoretical morphospace are geometric model parameters, different locations within the theoretical morphospace simply represent morphologies produced by the combination of different model coordinates along the dimensional axes. Given a theoretical morphospace, one can then determine which of the hypothetical forms seen within that morphospace have

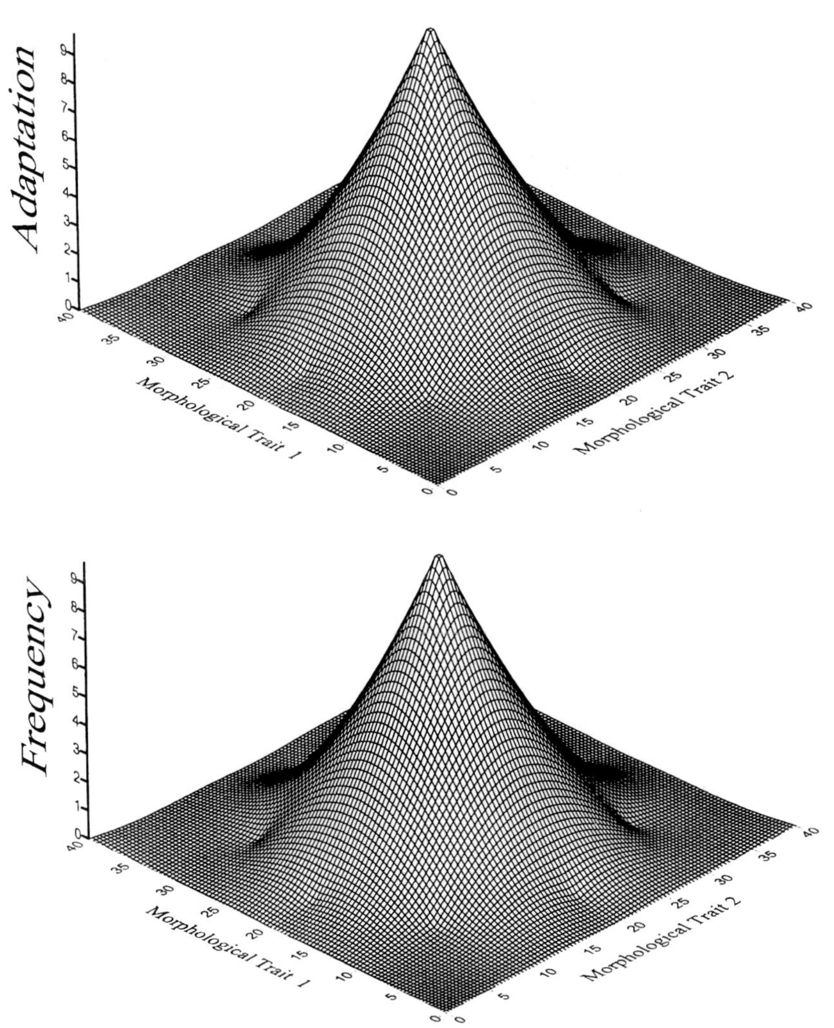

Figure 4.2. Contrasting the concept of an adaptive landscape (top figure) with that of a theoretical morphospace (bottom figure). The two morphological trait dimensions are the same in the two concepts. The vertical dimension, however, is the adaptive value of the different morphological combinations in the adaptive landscape. In the theoretical morphospace concept, the vertical dimension is the frequency of occurrence in nature of the different morphological combinations.

actually been produced in nature and those which, though geometrically possible, have not. The absence of actual biological forms within a theoretical morphospace does not necessarily mean that the hypothetically possible, but naturally nonexistent, morphologies are nonadaptive or have zero fitness (something that would be automatically assumed in an adaptive landscape). An alternative point of view could just as easily maintain that such hypothetical nonexistent morphologies might function perfectly well in nature, but that the process of evolution simply has not produced them yet. This point will be considered in more detail in Chapter 7.

A crucial concept shared by both theoretical morphospaces and adaptive landscapes (at least theoretically) is *their ability to specify nonexistent form*: e.g. Sewall Wright's possible but nevertheless nonexistent genotypes in the case of his fitness landscape and Dave Raup's possible but nevertheless nonexistent sea shells in his theoretical morphospace. And, although adaptive landscapes and theoretical morphospaces share the feature of hyperdimensionality, the dimensionality of a theoretical morphospace is always much less than that envisaged by Sewall Wright as necessary for an adaptive landscape large enough to encompass all possible combinations of genes present in life on Earth.

Another important feature of the dimensions of a theoretical morphospace is that they are defined without any reference to actual measurement data from existent form (we shall see an example of how this is done later in this chapter). The very ability of a theoretical morphospace to reveal nonexistent form is a function of the measurement-independent nature of the dimensions of the morphospace.

Finally, the concepts of an adaptive landscape and a theoretical morphospace converge if the dimension of 'degree of adaptation' or 'fitness' is somehow also mapped into a theoretical morphospace. Some examples of the convergence of these two concepts will be considered in Chapter 5.

Procedural phases of theoretical morphospace analyses

Consider again the room whose floor is covered with beautiful glass models of sea shells encountered at the beginning of this chapter. Now imagine that you and your family and friends spend an entire delightful week at the sea shore, relaxing and collecting sea shells as you walk along the shore. At the end of that week, you all take your collections and enter the room with the glass sea-shell models and carefully walk along each row of models, row after row, placing each real sea shell on top of the glass model that it most closely resembles. After all of your collected sea shells have been placed on the floor, you stand back and view the room. What you will see will be large piles of sea shells in some parts of the room, smaller piles of sea shells in other parts of the room, and some areas of floor where there are no sea shells whatsoever, even though the glass models are lying there. You are conducting a theoretical morphospace analysis of the evolution of actual sea-shell form in nature.

There are five procedural steps in conducting a theoretical morphospace analysis of actual form in nature (Fig. 4.3):

Step One: in order to construct a theoretical morphospace you will need to start with a geometric or mathematical model of morphology itself, or a mathematical model of the morphogenetic process that creates morphology. Both approaches are equally valid; choice of one or the other depends upon what you are most interested in (form simulation or growth simulation). In many cases creating such a model is not very difficult, it simply requires a little thought. An example of a simple model of form will be given in the next section of this chapter; for numerous additional examples see McGhee (1999).

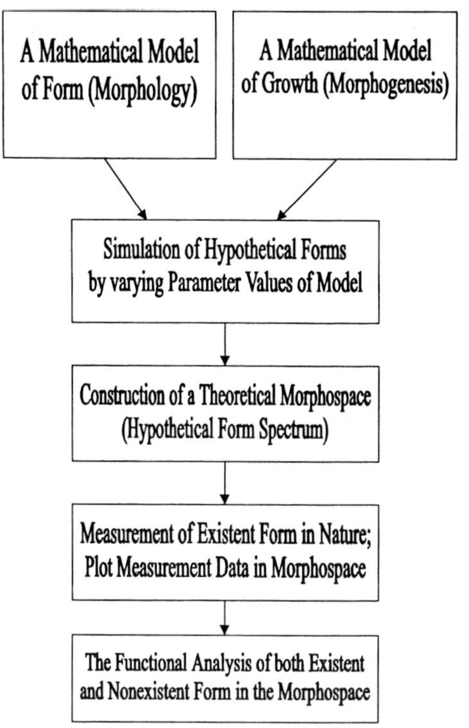

Figure 4.3. Flowchart of the procedural steps in a theoretical morphospace analysis.

Step Two: simulating the hypothetical morphologies that can be produced by the mathematical model. The actual simulations of form produced by different parameter combinations of the model can be as simple as line drawings, requiring nothing more complicated than protractors and compasses (and a pocket calculator for the calculations), to elegant graphics produced on your desk or laptop computer. In the next section of this chapter we shall examine some shell form simulations that were produced by a computer, but that could just as easily been drawn with a ruler and a compass.

Step Three: constructing a theoretical morphospace of hypothetical yet potentially existing morphologies. Once you have simulated a spectrum of hypothetical morphologies you can use the parameters of the mathematical model that produced them as the dimensions of your theoretical morphospace. Because the parameters of the model potentially determine the dimensionality of your morphospace, you want to use as few parameters as possible. A morphospace that has one million dimensions may be an interesting concept, but it is also an impractical one. Minimizing the dimensionality of the morphospace required to examine a particular group of organisms adequately is a major part of this phase of the analysis. Thus if your model has only three parameters, you can use them all to create a three-dimensional morphospace. If your model has eight or ten parameters, you probably will want to use fewer than that number in your final morphospace. Examine the spectrum of form produced by the model by systematically varying its parameter values. If variation in one or more of the parameters of the model produces small changes in form, relative to variation in other parameters of the model, then those parameters could be omitted in the construction of the final morphospace.

Step Four: measuring and plotting the spectrum of an existing form in the morphospace. Once you have constructed a theoretical morphospace of hypothetical form, you next will need to take measurements of the parameter values directly from actual organisms that fall within the hypothetical range of morphologies produced by your model. Obviously in this phase it is crucial to be able to obtain measurements of your model parameters from existing organisms. If your parameters are too abstract and unobtainable from actual biological form, then you must return to Step One and start over again with a different model.

It is in this phase of the analysis that you will begin to see which of your possible geometries or forms exist in nature, that have been produced in the evolution of life, and which equally possible geometries do not exist in nature. You will be able to see nonexistent form, in that you have before you a perfectly possible geometry in a simulated form that nevertheless has never been produced by any living organism. With enough data, you will be able to map out the regions of an existing and nonexistent form for a group of animals or plants within the theoretical morphospace.

Step Five: analyse the functional significance of both existing and nonexistent form in the morphospace. This is perhaps the most difficult, or labourious, phase of a theoretical morphospace analysis, yet is essential to determine whether the observed distribution of form within the morphospace is of adaptive significance. It is in this phase that the concepts of the adaptive landscape and the theoretical morphospace begin to converge.

The revelation of the spectrum of existing morphology within the realm of theoretically possible morphology does not in itself interpret the adaptive significance of the spectrum, but it does provide a powerful vehicle to facilitate the functional analysis of that spectrum of form. In most functional morphological studies, the morphologist concentrates on analysing the functional or bioengineering properties of the existing form that they have before them in their laboratory, or in the field. Sometimes this is an easy task, often is quite complicated. In theoretical morphospace analyses, however, you have the additional advantage of having both existing and nonexistent forms before you, not just the existing form. That is, you can not only generate and test functional hypotheses about the adaptive significance of a particular existing morphology, you can further test those hypotheses by applying them to the nonexistent morphologies within the morphospace.

In summary, the five steps of a theoretical morphospace analysis (Fig. 4.3) can be simply viewed as the creation of a morphospace, the exploration of a morphospace and the analysis of evolution within a morphospace. Now let us consider a real example of a theoretical morphospace.

Creating theoretical morphospaces: an example

In 1967 Dave Raup, the founder of the discipline of theoretical morphology, was interested in the evolution of shell form in ammonoids. Ammonoids are an extinct group of swimming cephalopod molluscs, related to the chambered nautilus that still survives in today's oceans.

The living chambered nautilus is but a pale shadow of the past glory of the swimming cephalopod molluscs, in that the oceans of the Palaeozoic and Mesozoic were filled with hundreds of their species and millions of their intricately coiled-shell individuals.

Raup considered the geometry of ammonoid shells and decided that the essence of their form could be simulated with a two-dimensional logarithmic spiral model. In polar coordinates, a two-dimensional logarithmic spiral has three geometric parameters: its radius r, its coiling angle φ, and its tangent angle α (Fig. 4.4), and is described by the exponential equation (4.1):

$$r_{\varphi} = r_0 \mathrm{e}^{(\cot \alpha)\varphi} \tag{4.1}$$

where r_0 is the magnitude of the initial radius you start with in the simulation, and r_{φ} is the magnitude of the radius at coiling angle φ. The variables r_{φ} and φ are constantly changing as the spiral grows, but the tangent angle α does not. This constancy of the tangent angle means that the logarithmic spiral has the additional geometric property of isometry; that is, the proportions of the spiral remain the same regardless of the size of the spiral.

Raup preferred to measure ratios of radii rather than tangent angles (it is easier!), thus he designed a new parameter to replace the tangent angle, which he named the 'whorl expansion rate', or W (see 4.2):

$$W = (r_{\varphi}/r_0)^{2\pi/\varphi}. \tag{4.2}$$

He then re-wrote equation (4.1) in the following fashion, in terms of his new parameter W as in (4.3):

$$r_{\varphi} = r_0 W^{\varphi/2\pi}. \tag{4.3}$$

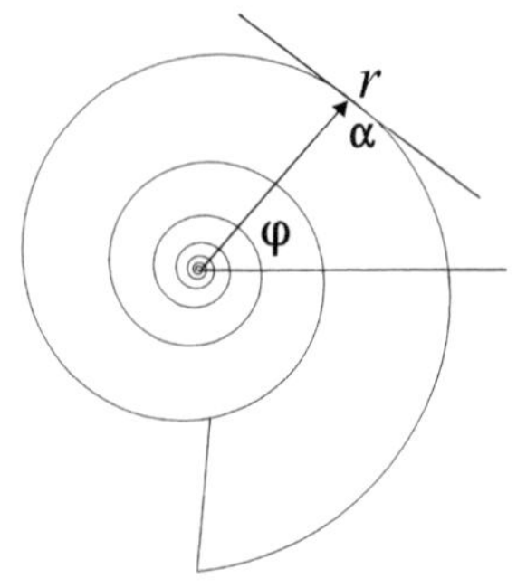

Figure 4.4. A logarithmic spiral, illustrating its three geometric parameters: radius r, coiling angle φ, and tangent angle α.

Examining a typical ammonoid shell, Raup decided that the essential aspects of the shell could be described by two logarithmic spirals: one spiral describing the outermost edge of the whorls of the shell, and a second spiral describing the innermost edge of the shell. He then designed a new parameter D, the distance from the coiling axis to the aperture, to measure the relative distance between the innermost and outermost spirals in the shell. The morphology of the ammonoid shell thus could be modelled by the twin equations (4.4) and (4.5):

$$\text{Outermost spiral } r_\varphi = r_0 W^{\varphi/2\pi}, \tag{4.4}$$

$$\text{Innermost spiral } r_\varphi = D(\text{outermost } r_\varphi). \tag{4.5}$$

Using these two equations, Raup wrote a brief computer program to simulate hypothetical ammonoid shells; that is, he wrote the program to simply draw the different spirals that could be produced by changing the parameter values of W and D in the mathematical model. The aperture of the shell lies between the innermost and outermost spirals of the shell, and Raup modelled the shell aperture to be a simple circle in the initial simulations. He then arranged these simulated shells into a two-dimensional theoretical morphospace of hypothetical ammonoid form, as illustrated in Figure 4.5, where the morphological dimensions of the morphospace are the parameters of the model, W and D.

In Figure 4.5 Raup has created a theoretical morphospace of hypothetical ammonoid shell form, the third step in a theoretical morphospace analysis (Fig. 4.3). All of the simulated shell forms are possible geometries for ammonoids to use as actual shells. If we use the traditional morphological terminology of ammonoid palaeontologists, the hypothetical shell forms illustrated in Figure 4.5 range, from the upper left in the morphospace to the lower right, from shell forms that are convolute to involute to advolute to evolute.

The various hypothetical ammonoid shells illustrated in Figure 4.5 were produced by a computer program, but they could just as well have been produced by using a ruler, a protractor and pocket calculator to perform the calculations. The computer simply does the job much faster; that is, the computer was used chiefly as a labour-saving device. The code that was used in producing the computer simulations shown in Figure 4.5 is given in Table 4.1.

What types of shell geometries did the ancient ammonoids actually evolve? The next step in a theoretical morphospace analysis is the actual measurement of an existing form in nature, and plotting those data in the

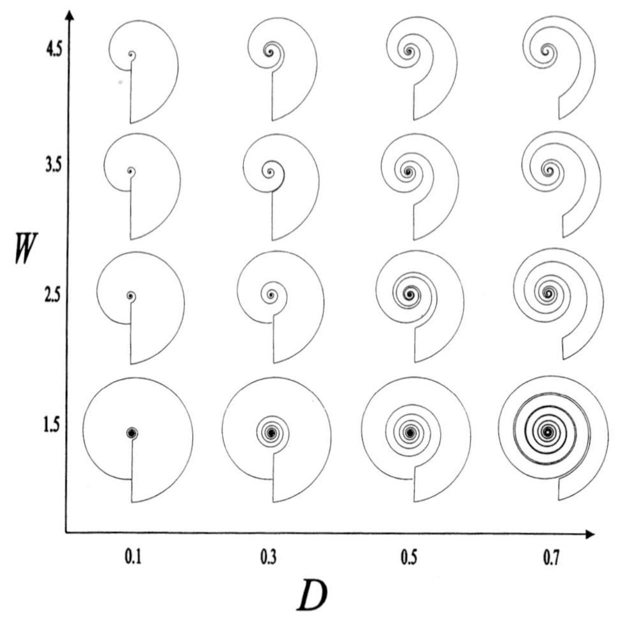

Figure 4.5. The theoretical morphospace of ammonoid form. The two morphological trait dimensions of the morphospace are the model parameters *W*, the whorl expansion rate of the shell, and *D*, the distance from the coiling axis to the aperture of the shell. These computer simulations were produced using the source code given in Table 4.1.

theoretical morphospace (Fig. 4.3). Raup deliberately designed his two model parameters *W* and *D* such that their actual values could be measured easily, and quickly, from an actual ammonoid shell (Fig. 4.6). He proceeded to measure the model parameters *W* and *D* on the actual shells of 405 ammonoid species, and to plot those measurements in the theoretical morphospace. He was interested in the frequency of occurrence of different shell geometries within the ammonoids: which shell forms were the most abundantly found among the ammonoids, which were rarer, and which were never found in ammonoids? Therefore he contoured the density of his data point distribution within the morphospace, as illustrated in Figure 4.7. The topographic high, or peak, in Figure 4.7 thus shows the most frequently occurring shell geometry in ammonoids, and the slopes away from the high point show shell geometries that are less abundant than shell geometries that are rarely found in ammonoids. Shell morphologies in the morphospace outside the outermost contour of the data density distribution are not found at all – they are nonexistent ammonoid shell forms.

Table 4.1. *Source code for producing the computer simulations given in Figure 4.5, written in Visual Basic 6.0. For use with a VB6 Form containing a PictureBox and three Command buttons: Run, Clear and Print*

```
Private Sub Command1_Click()
Dim OX(1 To 1000)
Dim OY(1 To 1000)
Dim IX(1 To 1000)
Dim IY(1 To 1000)
'Specify four initial parameter values (D, W, R0, PHI)
'Change these values as desired for other simulations
D = 0.5
W = 1.5
R0 = 1000
PHI = 1.5708
'Locate left edge of the simulation (LE)
t = 0.1
For j = 1 To 180
        x = R0 * Cos(PHI) * (W^(-PHI/6.2832))
        If (x > t) Then Exit For
        t = x
        PHI = PHI + (3.1416/180)
Next j
LE = t
PHI = 0
'Specify outer aperture margin (OMX, OMY)
OMX = R0
OMY = 0
'Specify inner aperture margin (IMX, IMY)
IMX = OMX * D
IMY = 0
'Locate bottom edge of simulation (BE)
PHI = 3.1416
t = 0.1
For j = 1 To 180
        Y = R0 * Sin(PHI) * (W^(-PHI/6.2832))
        If (Y > t) Then Exit For
        t = Y
        PHI = PHI + (3.1416/180)
Next j
BE = t
'Determine the length of the spiral (PHImax)
PHImax = (6.2832) * Log((200 * R0 * (1 - D))/(OMX - LE))/Log(W)
'Plot Outer Spiral (OX(j), OY(j)), reverse growth (anticlockwise)
N = PHImax * 180/3.1416 + 1
N = N/5
PHI = 0
Rout = OMX
OX(1) = Rout * Cos(PHI) * (W^(-PHI/6.2832)) - LE
OY(1) = Rout * Sin(PHI) * (W^(-PHI/6.2832)) - BE
        OX(1) = OX(1) * 2            'scale the plot (change as desired)
```

Table 4.1 (*cont.*)

```
        OY(1) = OY(1) * 2           'scale the plot (change as desired)
PHI = PHI + (3.1416/36)
For j = 2 To N
        OX(j) = Rout * Cos(PHI) * (W^(-PHI/6.2832)) - LE
        OY(j) = Rout * Sin(PHI) * (W^(-PHI/6.2832)) - BE
        OX(j) = OX(j) * 2           'scale the plot (change as desired)
        OY(j) = OY(j) * 2           'scale the plot (change as desired)
        PHI = PHI + (3.1416/36)
        Picture1.DrawWidth = 2
        Picture1.CurrentX = 3000
        Picture1.CurrentY = 6000
        If (W >= 1/D) Then
        Picture1.Line (OX(j-1) + 3000, 6000 - OY(j-1)) - (OX(j) + 3000, 6000 - _
  OY(J))
        Printer.Line (OX(j-1) + 3000, 6000 - OY(j-1)) - (OX(j) + 3000, 6000 - _
  OY(J))
        ElseIf (W < 1/D And PHI <= 6.2832 + 3.1416/36) Then
        Picture1.Line (OX(j-1) + 3000, 6000 - OY(j-1)) - (OX(j) + 3000, 6000 - _
  OY(J))
        Printer.Line (OX(j-1) + 3000, 6000 - OY(j-1)) - (OX(j) + 3000, 6000 - _
  OY(J))
        End If
Next j
PHI = PHI - (3.1416/36)
N = N + 100
'Plot Inner Spiral (IX(m), IY(m)), forward growth (clockwise)
N = N + 20
Rin = IMX
If (PHI < 0) Then PHI = 0
IX(1) = Rin * Cos(PHI) * (W^(-PHI/6.2832)) - LE
IY(1) = Rin * Sin(PHI) * (W^(-PHI/6.2832)) - BE
        IX(1) = IX(1) * 2           'scale the plot (change as desired)
        IY(1) = IY(1) * 2           'scale the plot (change as desired)
For m = 1 To N
        If (PHI < 0) Then PHI = 0
        IX(m+1) = Rin * Cos(PHI) * (W^(-PHI/6.2832)) - LE
        IY(m+1) = Rin * Sin(PHI) * (W^(-PHI/6.2832)) - BE
        IX(m+1) = IX(m+1) * 2           'scale the plot (change as desired)
        IY(m+1) = IY(m+1) * 2           'scale the plot (change as desired)
        If (PHI < 0.001) Then Exit For
        PHI = PHI - (3.1416/36)
        Picture1.Line (IX(m) + 3000, 6000 - IY(m)) - (IX(m+1) + 3000, 6000 - _
  IY(m+1))
        Printer.Line (IX(m) + 3000, 6000 - IY(m)) - (IX(m+1) + 3000, 6000 - _
  IY(m+1))
Next m
'Plot Line of Aperture
Picture1.Line (OX(1) + 3000, 6000 - OY(1)) - (IX(m) + 3000, 6000 - IY(m))
```

Table 4.1 (*cont.*)

```
Printer.Line (OX(1) + 3000, 6000 - OY(1)) - (IX(m) + 3000, 6000 - IY(m))
End Sub
Private Sub Command2_Click()
Picture1.Cls
End Sub
Private Sub Command3_Click()
Form1.PrintForm
End Sub
```

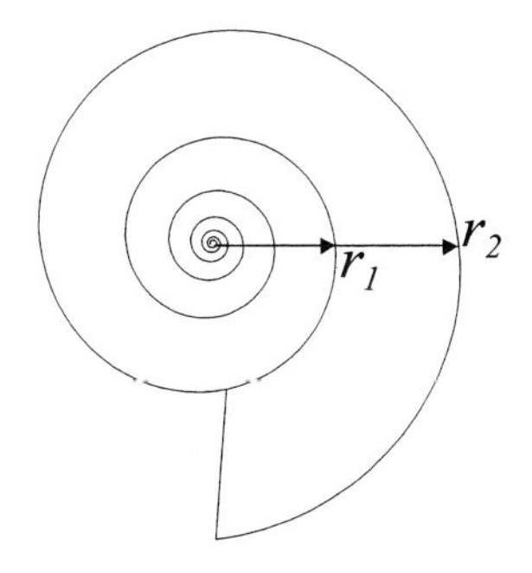

$$W = r_2 / r_1 \quad D = r_1 / r_2$$

Figure 4.6. Measurements needed to obtain the model parameters W and D on an ammonoid shell whose whorls do not overlap. For a discussion of measurement techniques on more complicated shell morphologies, see Raup (1967).

Note also the curved line, a hyperbola, within the morphospace that is labeled $W = 1/D$. This line is the *whorl-overlap boundary*; it delimits the region of the morphospace that contains shell forms that have whorls that overlap one another (the region to the left of the line) from the region that contains shell forms where the whorls do not touch one another (the region to the right of the line). Hypothetical ammonoids that fall exactly on the line within the morphospace have shell forms that have whorls that exactly touch one another, neither overlapping nor separating (see Fig. 4.5 for actual simulations of each of these morphological conditions). Of immediate interest is the fact that the huge majority of acutal ammonoids in nature occur to the left of the whorl-overlap boundary – very few ammonites possessed shells in which the whorls did not touch or overlap.

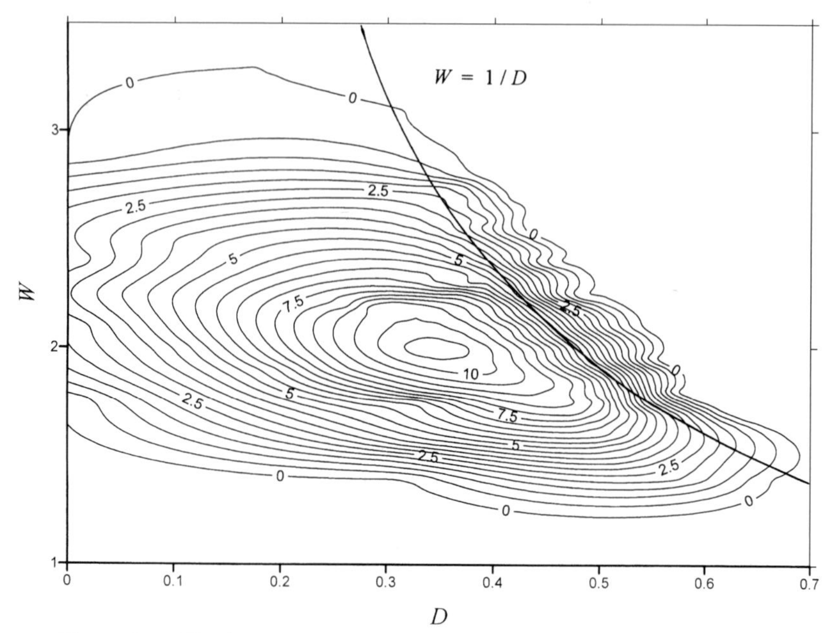

Figure 4.7. The frequency distribution of actual ammonoid morphologies found in 405 species, in the theoretical morphospace of hypothetical ammonoid form given in Figure 4.5. Note that the peak of the contours represents the most frequently occurring morphology in real ammonoids, and not the adaptive value of that morphology.
Source: Data from Raup (1967).

In Figure 4.7 Raup has completed the fourth step in a theoretical morphospace analysis, the exploration of the distribution of natural form within the morphospace. Note that the topographic high and lows in Figure 4.7 represents the frequency of occurrence of different shell forms in nature, and does not represent degrees of adaptation (see Fig. 4.2); thus Figure 4.7 is *not* an adaptive landscape.

The purpose of this section of the chapter has been to give an actual example of theoretical morphospace construction and to demonstrate that the process need not be complicated. The final step in a theoretical morphospace analysis is the functional analysis of both existing and nonexistent form within the morphospace (Fig. 4.3). It is often the most difficult step and, curiously, Raup himself did not complete it! But one of his graduate students did, and the results will be discussed in the next chapter.

5

Analysing the role of adaptive evolution in theoretical morphospaces

In studying the functional significance of the coiled shell, it is important to be able to analyze the types that do not occur in nature as well as those represented by actual species. Both digital and analog computers are useful in constructing accurate pictures of the types that do not occur.

Raup and Michelson (1965, p. 1294)

Functional analysis in theoretical morphospace

We saw in the last chapter that the five steps of a theoretical morphospace analysis (Fig. 4.3) can be summarized in three conceptual phases: the creation of a morphospace, the exploration of a morphospace, and the analysis of evolution within a morphospace. The analysis of evolution within a morphospace involves the functional analysis of the spectrum of both existent and nonexistent form within the morphospace, a spectrum that has been revealed in the first two phases of the analysis. The goal of functional analyses is to determine whether the observed distribution of form within the morphospace is indeed of adaptive significance, and it is in this phase of the analysis that the concepts of the adaptive landscape and the theoretical morphospace begin to converge.

In Chapter 4 we examined the process involved in creating a theoretical morphospace of hypothetical ammonoid morphologies, and the plotting of species of actual ammonoids within that morphospace, from the early work of Dave Raup, the founder of theoretical morphology. Raup did not complete the final step of the analysis – the analysis of the functional significance of the spectrum of existent and nonexistent ammonoid form – but his graduate student, John Chamberlain, did.

Adaptive significance of existing and nonexistent ammonoid form

Let us return to the frequency distribution of 405 ammonoid species within Raup's ammonoid morphospace that we considered in Chapter 4. The peak of the distribution, showing the most frequently occurring morphology found within actual ammonoids, occurs where the value of $D = 3.5$ and $W = 2.0$ (see Fig. 5.1). The frequency distribution is markedly asymmetrical, as can be seen in the pattern of the density contours. The distribution slopes sharply away on either side of an elongate ridge which trends from left to right across the figure and which is oriented subparallel to the D-axis of the morphospace. Thus variation in the parameter D in ammonoids is much greater than variation in W and the ammonoid values of the two parameters have a slight negative correlation. Note also the very steep slope of the morphological frequency distribution where the distribution abuts the whorl-overlap boundary in the morphospace (the hyperbola labelled $W = 1/D$ in Fig. 5.1).

What is the functional significance of the observed ammonoid frequency distribution within the morphospace? As noted by Raup (1967), there might not be any at all, in that the argument could be made that the frequency distribution of ammonoid morphologies that did evolve in time may simply represent chance. That is, if the ammonoids had had more time to evolve new morphologies (and not have met extinction 65 million years ago), that perhaps the empty regions of the morphospace seen in Figure 5.1 might eventually have been filled. We shall examine this idea in more detail in Chapter 7, when we consider the concept of evolutionary constraint within morphospace.

The alternative argument from natural selection theory is, of course, that there is an adaptive advantage associated with the morphologies that the ammonoids have evolved, and an adaptive disadvantage with those which have not. Raup (1967) speculated that the observed frequency distribution of ammonoid morphologies within the morphospace does not represent the optimization of any single functional aspect of shell form, but rather that the geometries present in the occupied region of the morphospace represented shell forms which minimize several different functional problems faced by the ammonoids. Ammonoids were active swimmers, and thus the shell forms that they did evolve might have something to do with streamlining and orientation stability constraints associated with their swimming mode of life. These swimming constraints, particularly the need for streamlining, might partially explain why the great majority of ammonoids have shells in which the whorls contact and overlap one another, producing a solid disk with no open spaces between the whorls. The location of the ammonoid frequency distribution to the left of the whorl-overlap boundary in Figure 5.1 could also be due, at least in part, to bioeconomical constraints in efficient secretion of shell material by the animals (overlapped whorls cost less shell material) and due to increased strength in these types of shell (shells with overlapped whorls are stronger, and can better resist crushing by predators).

The theoretical suggestions of Raup (1967) concerning the adaptive significance of ammonoid shells have been corroborated and expanded by the experimental work of his graduate student, John Chamberlain (1976, 1981). Using a series of shell models, Chamberlain (1976) experimentally determined the drag coefficients for a variety of different shell geometries. Not unexpectedly, he found that shells with whorl overlap have much lower drag coefficients than those without. I have here converted Chamberlain's drag coefficients to swimming-efficiency coefficients in order to create a topographical space where the maximum swimming-efficiency coefficient (a topographic high) represents the minimum drag coefficient, and vice versa (Fig. 5.1).

For shells with whorl overlap, Chamberlain found that two regions of maximum swimming-efficiency coefficients exist within the morphospace: one where $W = 1.5$ and $D = 0.05$ to 0.13, and another where $W = 1.9$ and $D = 0.35$ to 4.2 (Fig. 5.1). These adaptive peaks have swimming-efficiency coefficients greater than 70. The first adaptive peak corresponds to shells that are highly involute (computer simulation on the left), and the second to shells with only moderate whorl overlap and a wide umbilicus (computer simulation in the middle). The computer-simulated evolute shell shown on the right lies within the region of the morphospace where the whorls of the shell do not touch one another, and it has a very low swimming-efficiency coefficient (around 10).

The swimming-efficiency-coefficient surface shown in Figure 5.1 is, in essence, an adaptive surface. It shows the degree of adaptation for each potential shell geometry within the morphospace, where the degree

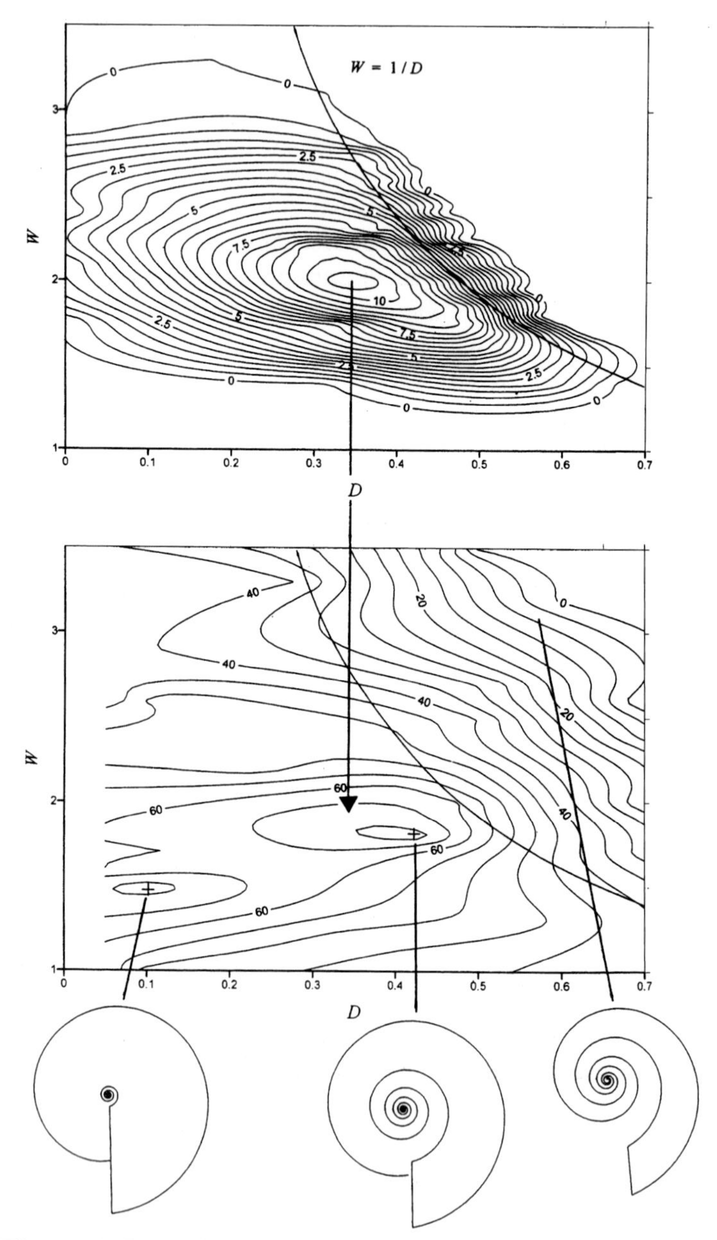

Figure 5.1. Comparison of the contoured frequency distribution of 405 species of actual ammonoid forms in theoretical morphospace (top figure) with the contoured distribution of swimming-efficiency coefficients within theoretical morphospace (bottom figure).
Source: Swimming-efficiency data from Chamberlain (1981).

of adaptation is measured in terms of the degree of streamlining that an ammonoid would experience while swimming with any particular shell geometry.

Comparing the topology of the frequency distribution of real ammonoid shells within the morphospace and the topology of the adaptive surface of maximum swimming-efficiency reveals a surprise – only one of the adaptive peaks is occupied (Fig. 5.1)! The maximum peak of most abundant ammonoid shell forms in nature is very close indeed to the experimentally-determined maximum swimming-efficiency peak around $W = 1.9$ and $D = 0.35$ to 4.2 in the morphospace (as shown by the arrow in Fig. 5.1). Thus Chamberlain was able to demonstrate that a major adaptive determinant in the evolution of shell form in ammonoids is maximizing streamlining efficiency in swimming.

Why is the swimming-efficiency adaptive peak on the left unoccupied? For many years this was a mystery. One possibility is that the ammonoids were *unable* to develop morphologies in this region of the morphospace – the concepts of developmental and phylogenetic constraint, which we shall examine in more detail in Chapter 7. Alternatively, there may be some additional functional property, in addition to streamlining efficiency, that renders shells with the geometries found in this region of the morphospace maladaptive. Several suggestions include maladaptive bouyancy stability of these shells, or simply that there is too little space in the outer whorl of these shells to contain the body of the ammonoid!

Now, some four decades after Raup's original study, we finally have the answer to this mystery – the peak is indeed occupied! Saunders, Work and Nikolaeva (2004) point out that Raup (1967) used a 1957 database of ammonoid systematics, the best available at the time, that is now vastly out of date. In the past half-century, 588 new genera of Palaeozoic ammonoids have been discovered and described. Saunders, Work and Nikolaeva (2004) took measurements from 597 species of these newly described ammonoids, plotted them in Raup's theoretical morphospace of ammonoid form, the results of which are shown in Figure 5.2. These new data reveal an ammonoid-form frequency distribution that is much more rugged than that originally seen by Raup (Fig. 5.1). Saunders, Work and Nikolaeva (2004) discern three main peaks in shell frequency, instead of one, located along a ridge centred at $W = 1.9$. All ammonoid shell forms located along this ridge have swimming-efficiency coefficients greater than 60 (Fig. 5.2). Most importantly, these new data reveal that the Palaeozoic ammonoids did evolve forms very close to the swimming-efficiency adaptive peak on the left

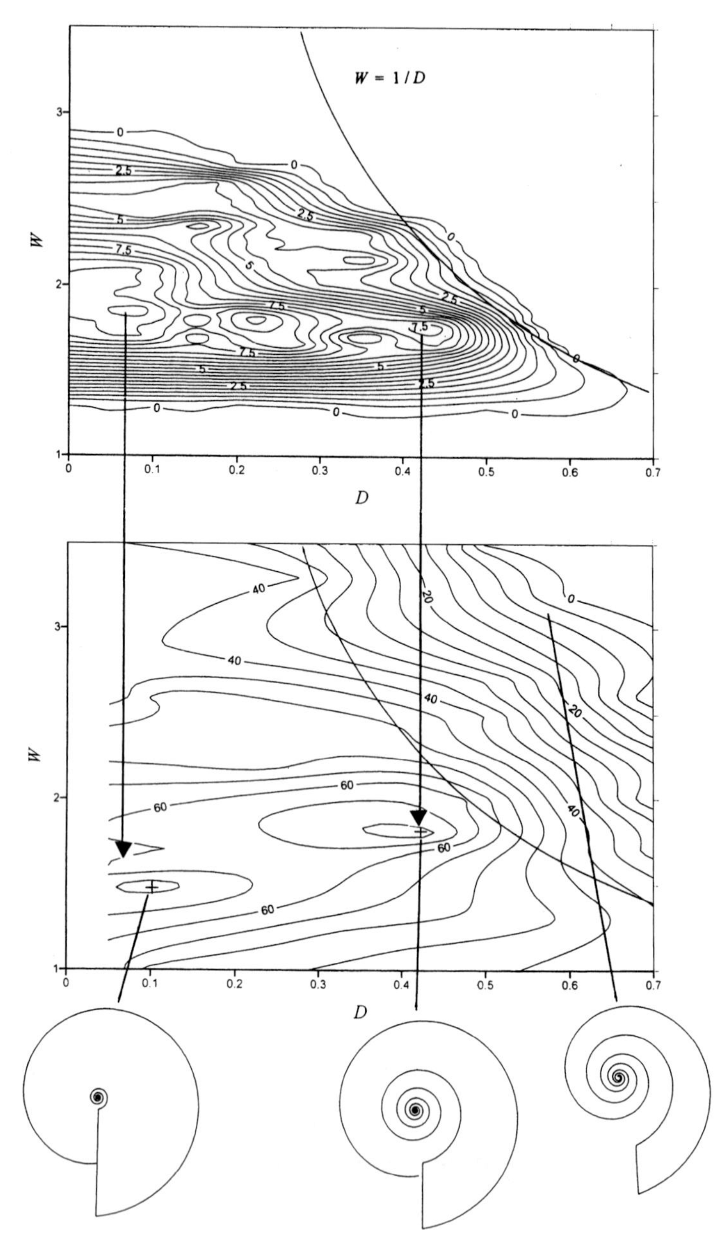

Figure 5.2. Comparison of the contoured frequency distribution of 597 species of Palaeozoic ammonoids. Measured by Saunders, Work and Nikolaeva (2004), in theoretical morphospace (top figure) with the contoured distribution of swimming-efficiency coefficients within theoretical morphospace (bottom figure). In the bottom figure, both of the two adaptive peaks of maximum swimming-efficiency coefficients are shown to have been occupied by the Palaeozoic ammonoids.
Source: Ammonoid data from Saunders, Work and Nikolaeva (2004).

(Fig. 5.2) that for so many years was thought to have been an empty region of morphospace (Fig. 5.1). We have now completed the last phase of a theoretical morphospace analysis of the adaptive significance of both existent and nonexistent shell form in the extinct ammonoids. We have seen that the most significant adaptive determinant of shell geometry in the ancient ammonoids was shell streamlining for efficient swimming (Figs. 5.1 and 5.2).

In this discussion, I purposefully did not skip directly to the recent study of Saunders, Work and Nikolaeva (2004) in order to illustrate the predictive power of the techniques of theoretical morphospace analysis. The adaptive surface in Figure 5.1 would predict, under the expectations of the theory of natural selection, that ammonoids should have evolved forms in the apparently empty region of the morphospace. Empty morphospace in and of itself leads to the concept of evolutionary constraint, which we shall examine in detail in Chapter 7. Empty morphospace that can be demonstrated to be an adaptive peak is even more interesting and anomalous. In the present case, the anomaly turns out to have been simply due to an inadequate database. In Chapter 7 we shall see that it is indeed theoretically possible for adaptive peaks to remain unoccupied, and to consider the potential causes of such a phenomenon.

The crucial point to be seen in Figures 5.1 and 5.2 is that they give an actual adaptive surface and a real frequency distribution of ammonoid shell forms that have evolved in nature. They are not merely conceptual models, heuristic devices for thinking about evolution. They give an actual analysis of the adaptive significance of morphology that has been evolved by an actual group of animals.

Adaptive significance of existing and nonexistent brachiopod form

The brachiopods, or lampshells, are another group of organisms that have been analysed for adaptation by using the techniques of theoretical morphology. Brachiopods are marine animals that superficially resemble clams, but in fact are not molluscs at all. They belong to a group of animals known as the lophophorates, as all of these organisms use delicate tentacular feeding organs known as lophophores to strain food particles out of the surrounding sea water; that is, they are filter feeders. Although brachiopods are still present in today's oceans, they were much more abundant in the past and are often referred to as the 'shellfish of the Palaeozoic' by palaeontologists.

A theoretical morphospace of brachiopod shell form can be created by using the same two geometric parameters that Raup devised for the swimming ammonoids in the previous chapter: the whorl expansion rate of the shell, *W*, and the distance of the aperture of the shell from the coiling axis, *D*. The major difference between the shells of ammonoids and brachiopods is that the brachiopods have two shells, or valves, that are articulated together (much like clams) instead of just one shell, like ammonoids. Anatomically, the two valves in a brachiopod shell are termed the dorsal, or upper valve, and the ventral, or lower valve. Thus, in order to characterize the geometry of a brachiopod shell, we have to specify a separate *W* value for each of two valves in the shell (Fig. 5.3). Brachiopods have shells that are very flat compared to the highly coiled shells found in ammonoids, thus they have very high whorl expansion rates and the dimensions of the morphospace shown in Figure 5.3 are scaled in logarithms, or orders of magnitude, of *W*.

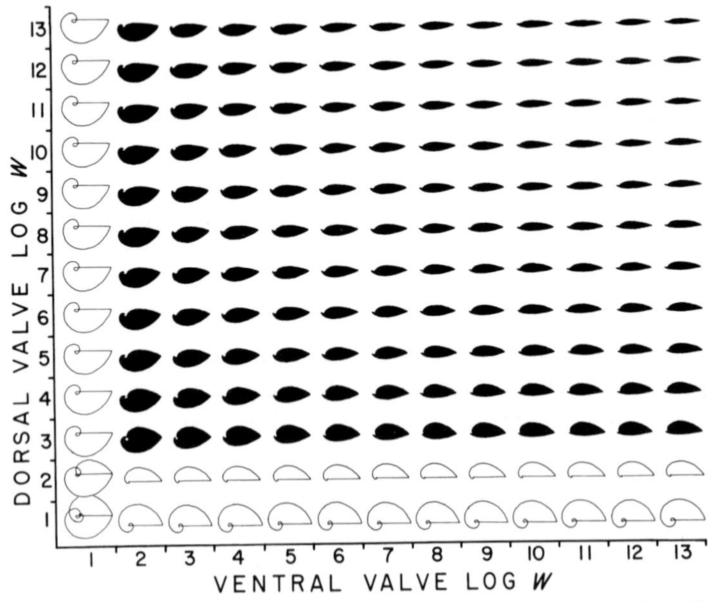

Figure 5.3. A theoretical morphospace of hypothetical brachiopod shells. The two morphological-trait dimensions are the dorsal and ventral whorl expansion rates, *W*, of the two valves that comprise the shell.
Source: From McGhee (1980a).

The reader will also note something else unusual in Figure 5.3. One vertical row of ventral valves is shown on the left in the morphospace, and two horizontal rows of dorsal valves are shown at the bottom, that are not articulated with their corresponding opposite valve to make a complete brachiopod shell (like all the other shells shown in the morphospace). Why not? This is because the single valves shown have whorl overlap, and *whorl overlap must be absent in both valves in order to articulate those valves together to form a functioning hinged bivalved shell.* The absence of shells with functioning hinged valves in this region of the morphospace is an example of *geometric constraint* in evolution; that is, it is geometrically impossible to created hinged bivalved shells in this region of the morphospace. We shall consider the concept of geometric constraint in evolution in more detail in Chapter 7.

In the case of brachiopod shells, the values of W and D in both dorsal and ventral valves must meet the geometric condition that the magnitude of $W < 1/D$. Typically, brachiopod shells have dorsal values of D of around 0.01 and ventral D values of around 0.1, which mean that typical brachiopods cannot produce shells with ventral valves with $\log W < 2$ and dorsal valves with $\log W < 3$ (Fig. 5.3). This region of the morphospace is empty of brachiopods, not due to low or zero adaptive potential, but due to the fact that it is impossible to form hinged bivalved shells at all in this region of the morphospace. To illustrate this impossibility graphically, two simulations are shown in the lower left corner of the morphospace (Fig. 5.3). Inescapable geometry would require the valves of the hypothetical shells to interpenetrate one another, to occupy the same space at the same time, which is impossible.

In Figure 5.3 the first phase of a theoretical morphospace analysis of shell morphology in biconvex brachiopods has been completed. To complete the second phase, measurements were taken from the shells of 324 species of actual brachiopods, and the frequency distribution of those shells was mapped into the morphospace (Fig. 5.4). The peak, or most frequently used, shell form in the examined brachiopods has a dorsal-valve $\log W = 5$, and a ventral-valve $\log W = 3$, and is thus inequivalved; that is, the shells have flatter dorsal valves than ventral. The frequency distribution of morphologies is positively skewed, with the majority of brachiopod shells occurring in regions of low W magnitudes, but with the distribution sloping outwards to the high magnitude regions of $\log W$ values, ranging from 10 to 12. As we knew had to be the case, no biconvex brachiopods are found with very low magnitudes of $\log W$, such as one to two, because shells in this region of the morphospace are geometrically impossible (Fig. 5.3).

Figure 5.4 gives the frequency distribution of actual brachiopod shell morphologies within the spectrum of geometrically possible shell forms (Fig. 5.3) that have been produced by nature in the long evolution of the group. The last phase of the analysis is entered when we ask: what, if anything, is the functional significance of this frequency distribution? Actual brachiopods occupy only a limited region within the theoretical continuum of potential shell geometries, that is, there exist possible geometries that nevertheless cannot be found among brachiopods in nature (compare Figs. 5.3 and 5.4). Is this chance? Or is there a functional reason for the evolution of so many brachiopod shells in the lower-left region of the morphospace (Fig. 5.4)?

Brachiopods are marine filter feeders. The animals pump outside water into their shells, where any food material present in the water is filtered out by the tentacles of the lophophore, and then the filtered water is pumped back out of the shell. The larger the brachiopod's lophophore, the more water it can filter for food. The lophophore, however, has to be contained within the protective shell of the animal. Although brachiopods have evolved complex ways to fold up their lophophores in order to pack big lophophores in small shells, there is still a limit to what can be accomplished by folding. The best way to create space within the shell for bigger lophophores is to grow a shell that has a large internal volume relative to its external surface area. The one three-dimensional geometry that has the smallest surface area, and largest volume relative to its surface area, is the sphere.

Figure 5.4 illustrates the contoured frequency distribution of volume-to-surface-area ratios for shell forms within the morphospace. The peak of maximum volume-to-surface-area ratios is located in the lower left region of the morphospace; that is, shell geometries in this region of the morphospace have the largest volumes and smallest surface areas of any shells in the morphospace. Comparing the topology of the frequency distribution of real brachiopod shells within the morphospace and the topology of the adaptive surface of volume-to-surface-area ratios (Fig. 5.4) reveals a close correspondence between the two topologies. The most abundant shell geometries found in the real brachiopods have volume-to-surface-area ratios of 10 or greater, and the frequency of occurrence in nature of real brachiopod shells with lower and lower ratios of volume-to-surface-area becomes less and less (Fig. 5.4). The fact that the frequency distribution peak is not exactly centred on the maximum

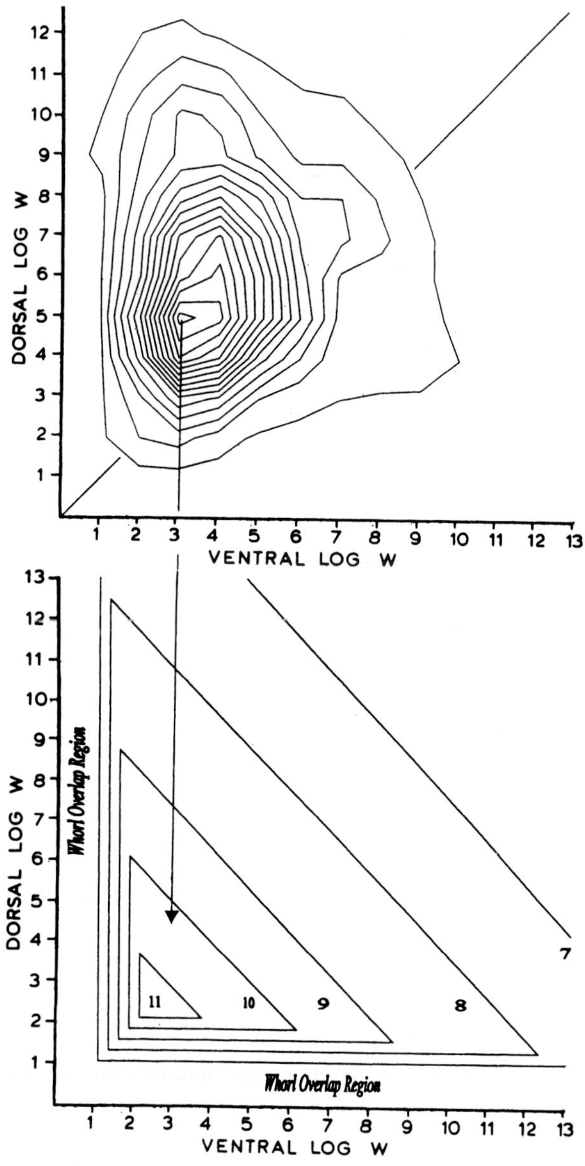

Figure 5.4. Comparison of the contoured frequency distribution of 324 species of actual brachiopod shell forms in theoretical morphospace (top figure) with the contoured frequency distribution of volume-to-surface-area ratios for shell forms within the morphospace (bottom figure). In the bottom figure, an adaptive peak of maximum volume to minimum surface-area can be shown to exist, which is very close to the frequency of form peak seen in actual brachiopods.
Source: Data from McGhee (1980a).

volume-to-surface-area ratio peak of 11, but is located slightly down-slope, may indicate the additional geometric constraint effects of shell hingeing in the real brachiopods (McGhee, 1980a).

Thus a major adaptive determinant in the evolution of shell form in brachiopods is maximizing internal volume; that is, brachiopods evolved shells to approximate a spherical form as closely as possible within specific geometric constraints (McGhee, 1980a). As in our considerations of the evolution of ammonoid form, the crucial point to be seen in Figure 5.4 is that it gives a real adaptive surface and a real frequency distribution of brachiopod shell forms that have evolved in nature. It is not merely a conceptual model, a heuristic device for thinking about evolution. It is an actual analysis of the adaptive significance of morphology that has been evolved by an actual group of animals.

Adaptive significance of existing and nonexistent plant form

Thus far we have considered examples of theoretical morphospace analyses of organic form in marine animals that swim, and marine animals that are sessile on the sea bottom. Life is not confined to the oceans, however, and not all multicellular life is animal. Numerous innovative analyses of the adaptive significance of terrestrial plant forms in theoretical morphospace have been conducted by Karl Niklas and his colleagues (Niklas and Kerchner, 1984; Ellison and Niklas, 1988; Niklas, 1986, 1997a, 1997b, 2004).

Figure 5.5 gives just one example of several theoretical morphospaces of plant form created by Niklas and his colleagues. The hypothetical plant forms shown in Figure 5.5 were computer simulated using a geometric model of plant form that has four parameters, two probabilistic and two geometric. The first two parameters are the probability that a branch will continue to grow once formed, and the probability of branching of a leader branch, a branch that has grown for at least two internodes. The second two parameters are the branch angle, ϕ, that a branch makes with the main plant axis when it forms, and the leader-internode ratio, *LIR*, which is the ratio of the lengths of the successive growth internodes of the branch (Ellison and Niklas, 1988). Note that hypothetical plant forms in the upper-left region of the morphospace are spindly and thin, with few branches,

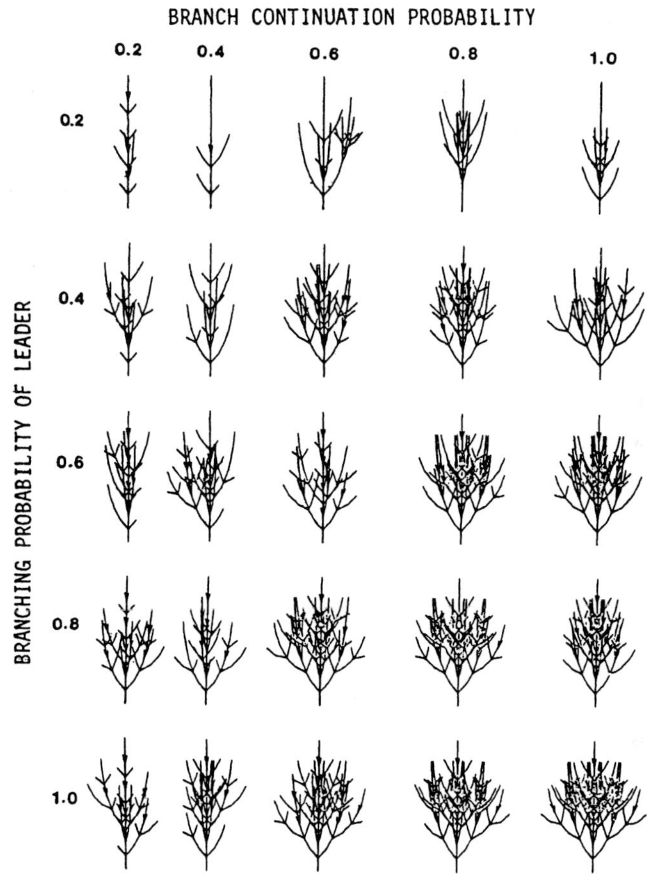

Figure 5.5. A theoretical morphospace of hypothetical plants. The two morphological-trait dimensions are the probability that a branch will continue to grow, versus the probability that the leader branch will bifurcate. Two other morphological parameters in the model are here held constant (i.e. $\phi = 45°$ and $LIR = 1.0$; see text for discussion).
Source: From Ellison and Niklas (1988). Copyright © 1988 by the *American Journal of Botany* and reprinted with the permission of the publisher.

whereas hypothetical plant forms in the lower-right region of the morphospace are bushy and thick (Fig. 5.5). Various hypothetical intermediate plant morphologies can seen within the morphospace, trending from the upper-left to the lower-right in the figure.

For each of the hypothetical plant geometries illustrated in Figure 5.5 it is possible to calculate the *light-interception efficiency* of that geometry. Plants are photoautotrophic organisms – they harvest *light*, rather plankton or nutrients like the filter-feeding brachio- ered in the previous section. Therefore it is of major lants to have maximum photosynthetic surface areas ction of light from the sun. But the *efficiency* with which plants collect light is not just a simple function of total photosynthetic surface area of the plant. They must also possess geometries that have *minimum shading* of the photosynthetic surface areas of the plant by other parts of the plant itself. The trick is to arrange the various branches of the plant at different angles and orientations from one another so as to minimize the amount by which the various branches shade one another, and thus cut off the the sunlight from the photosynthetic surfaces of the shaded branch.

Six different permutations of the model parameters, and the adaptive peaks and valleys of differing light-interception efficiencies of hypothetical plant geometries produced by those geometric permutations, are illustrated in Figure 5.6. The two horizontal dimensions of the three-dimensional block diagrams shown are the morphological dimensions of the theoretical morphospace given in in Figure 5.5, whereas the vertical dimension of the block diagrams is the light-interception efficiency of the plant geometries at the morphological coordinates within the theoretical morphospace. In the three block diagrams on the right, the branching angle ϕ is equal to 30°, and in the three block diagrams on the left ϕ is equal to 45°. In the two block diagrams at the top, the leader-internode ratio *LIR* is equal to 1.0, in the two middle block diagrams *LIR* is equal to 0.75, and in the two bottom block diagrams *LIR* is equal to 0.50. Computer simulated plant forms from selected regions in the morphospace are also given (see also Fig. 5.5).

Figure 5.6 is an actual adaptive landscape of terrestrial plant form, within the limits of the model that produced the theoretical morphospace in Figure 5.5. Note that the altitudes of the adaptive peaks within the landscapes become progressively higher, from the bottom landscapes to the top landscapes within the figure, in terms of the magnitude of the light-interception efficiency index values found in each landscape. This increase in the altitude of the adaptive peaks is a function of the increase in the parameter *LIR* in the theoretical morphospace; as the value of the leader-internode ratio increases, the branches progressively become longer and the total plant becomes taller (see simulations in Fig. 5.6). Note that, in general, hypothetical plants with high probabilities of branching, and continued growth of those branches, are in the adaptive valleys of minimum light-interception efficiency.

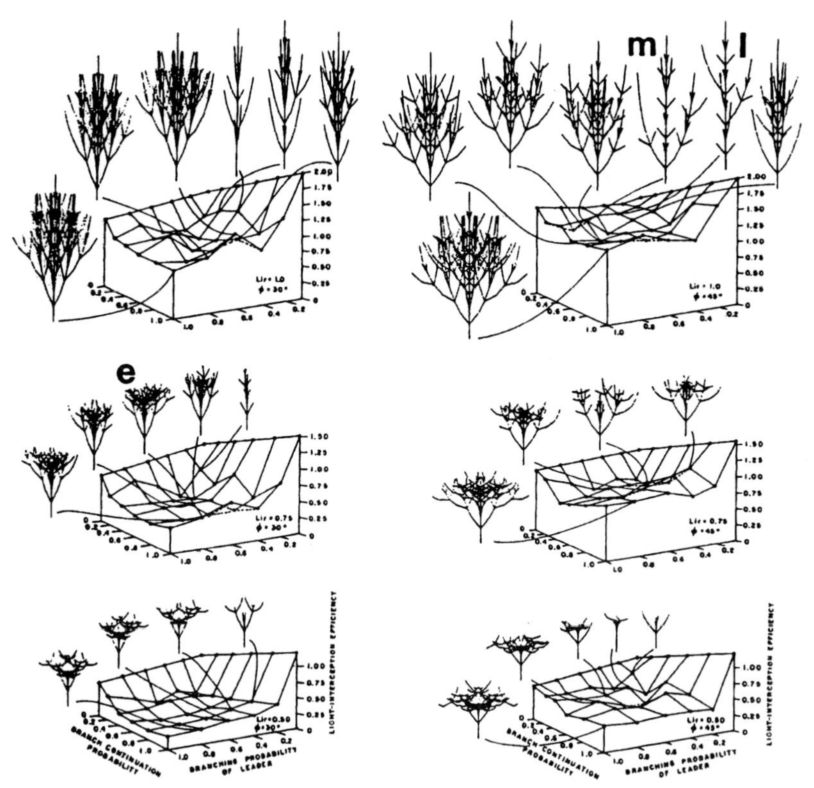

Figure 5.6. Adaptive peaks and valleys in the theoretical morphospace of hypothetical plants given in Figure 5.5. The adaptive peaks represent maximum light-interception efficiency by plant geometries in that region of the morphospace. Computer simulated plant forms from selected regions in the morphospace are indicated. Hypothetical plant forms that correspond to actually occurring plant forms in *Salicornia europaea* during different successional stages are indicated by letters; **e** indicates an early successional morphology, **m** a middle successional morphology, **l** a late successional morphology.
Source: From Ellison and Niklas (1988). Copyright © 1988 by the *American Journal of Botany* and reprinted with the permission of the publisher.

One major limitation of the model producing the theoretical morphospace in Figure 5.5 is that the plants have no leaves! Adding the geometries and orientations of leaves to the branching model would vastly increase its complexity. Yet still there exist many plants, both fossil and living, that do indeed have no leaves and that are photosynthetically active over the entire surface of the plant. One such plant is the living succulent *Salicornia europaea*, found in the salt marshes of

North America and Europe. Ellison and Niklas (1988) were interested in the phenomenon of *ecological succession*, the sequential replacement of one species by another following an ecological disruption of a plant community, and its subsequent recovery to its initial state. Following the disruption of a salt marsh community during a storm tide, *Salicornia europaea* is often the first colonizer in the disrupted area, but is eventually replaced by the return of the dominant marsh grasses. In contrast to the usual ecological succession studies, which focus on the sequential species replacements, Ellison and Niklas (1988) were interested in morphological changes that occurred in the *Salicornia europaea* plants themselves during the succession process. As *Salicornia europaea* is leafless, its morphology can be accurately modelled by the branching model alone.

Measurements taken from actual *Salicornia europaea* plants, obtained during different successional stages, were plotted within the theoretical morphospace by Ellison and Niklas (1988). The position of the early, middle and late successional plant morphologies are given in Figure 5.6. The early successional plants are quite bushy, with many branches having small internodes (see the middle-left adaptive landscape in Fig. 5.6). In contrast, middle and late successional plants are elongated, with many fewer branches, and with longer internodes (see upper-right adaptive landscape in Fig. 5.6). Ellison and Niklas (1988) have argued that these morphological changes – and the movement of *Salicornia europaea* plant morphologies across the adaptive landscapes in Figure 5.6 – are directly related to the reduction in light availability to *Salicornia europaea* during the successional process. As more and more individuals of the dominant marsh grasses return to the disturbed area, they progressively shade out the colonizer *Salicornia europaea* individuals. Early successional individuals of *Salicornia europaea* have relatively low light-interception efficiencies, but produce many seeds on their numerously-branched bushy forms. Thus they concentrate on reproductive potential in the absence of competition for light. Following the return of the marsh grasses, light becomes a limiting resource, and *Salicornia europaea* individuals produce geometries that have much higher light-interception efficiencies (Ellison and Niklas, 1988).

One final example from the work of Karl Niklas addresses the theoretical morphology of leaf shape and orientation in terrestrial plants. Consider the helicospiral arrangement of leaves along a branch: number the lowest leaf as the first leaf, the next leaf higher up along the branch as the second leaf, and so on. The second leaf on the branch is usually oriented at an angle to the first leaf; that is, the second leaf is not directly above the position of the first leaf on the branch. Farther along the branch, however, we usually encounter a leaf, say the nth leaf, which is directly above the position of the first leaf. We can then use the first leaf and the nth leaf to define a *phyllotactic period*, a period that has two descriptive components: the number of leaves in the period (in this case, n) and the number of whorls or windings that the leaves make along the branch in the period, which we can designate as w. Empirical examination of enormous numbers of plants reveals that both n and w are numbers in a *Fibonacci sequence*.

What is a Fibonacci sequence? The Fibonacci sequence is the mathematical sequence of the Italian mathematician Leonardo Pisano ('Fibonacci') who, in addition to introducing into Europe the numericals we today call 'arabic', was concerned with breeding rabbits in the year 1202. The solution to the problem of 'how many pairs of rabbits do I have?' with each successive reproductive cycle turns out to be the Fibonacci sequence, in which each number in the sequence is the sum of the two previous numbers: 1, 1, 2, 3, 5, 8, 13, 21, 34, 55, 89, 144, 233, . . . , and so on, where any given number, a, at position p in the sequence is:

$$a_{p+1} = a_{p-1} + a_p$$

where p is an integer of value of two or larger.

In the case of leaf arrangements, the values of w and n in many actual plants are two numbers in the Fibonacci sequence that are not immediately adjacent to each other, but rather are separated by a single number within the sequence; that is:

$$w/n = 1/2, 1/3, 2/5, 3/8, 5/13, 8/21, \ldots, a_p/a_{p+2}.$$

This relationship is known as the *phyllotactic fraction*, and the ratio of w/n for higher phyllotactic fractions, such as 8/21, 21/55, 34/89, and so on, results in each leaf being arranged around the branch at an angle of 137.5° to the previous leaf. This angle, 137.5°, is thus also known as the *Fibonacci angle*.

It is generally assumed that leaves are arranged in Fibonacci angles in order to minimize the shading of one leaf by another along the branch, and thus to maximize the total photosynthetic area exposed to the sun of the combined leaves of the plant. Using the analytical techniques of theoretical morphology, Niklas (1997b) was able to show that this is not always the case. In Figure 5.7 is given the *phyllotactic morphospace* of Niklas (1997b), which illustrates the combined geometric effect of

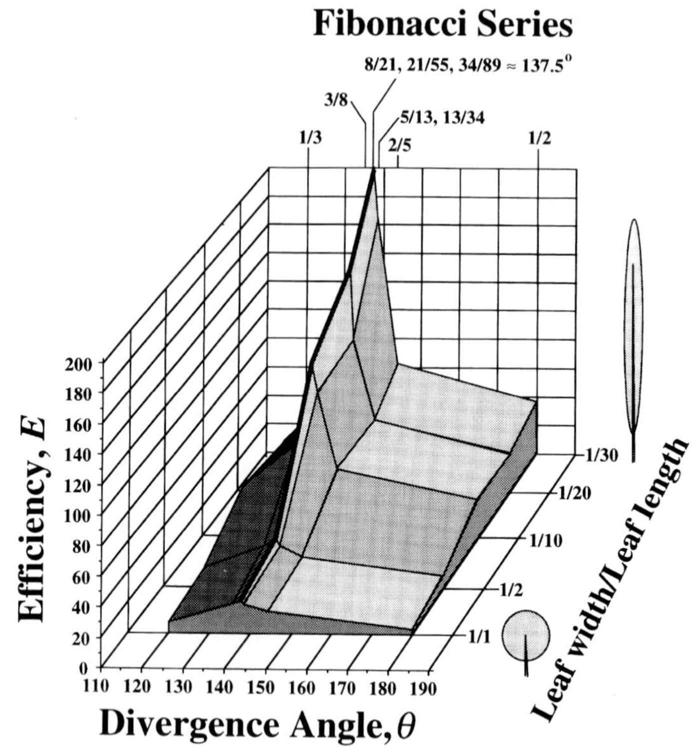

Figure 5.7. The *phyllotactic morphospace* of Niklas (1997b). The two morphological-trait dimensions are the width/length ratio of leaves (y-axis of the morphospace), and the divergence angle between leaves arranged along a branch (x-axis). The vertical axis is the light-interception efficiency (z-axis) of the geometric permutations of leaf shape and arrangement in the morphospace. At the top of the figure are given the various leaf-divergence angles produced by a series of phyllotactic fractions; note that the higher fractions converge on the Fibonacci angle of 137.5°. Narrow, slender leaves arranged in Fibonacci angles along branches have the highest light-interception efficiencies; leaves with a nearly circular outline have much lower efficiencies, even arranged in Fibonacci angles, due to overlapping leaf outlines leading to shading.
Source: Artwork courtesy of K. J. Niklas. From Niklas (1997b); copyright © 1997 by the University of Chicago Press and reprinted with the permission of the publisher.

both leaf arrangement and leaf shape on light-interception efficiency. Using this morphospace of hypothetical leaf geometries, Niklas (1997b) demonstrated that leaves arranged in Fibonacci angles to one another along a branch do indeed produce the maximum light-interception efficiency – but only if those leaves are long and slender (Fig. 5.7).

In contrast, hypothetical plants with circular leaves may have leaf arrangements which deviate significantly from the Fibonacci angle, as leaf divergence angles of 137.5° convey only a slight increase in light-interception efficiency (Fig. 5.7).

In summary, both Figure 5.6 and Figure 5.7 illustrate actual adaptive landscapes, created using the analytical techniques of theoretical morphology. The theoretical morphospace in Figure 5.6 is one for branching geometries, and in Figure 5.7 is one for leaf geometries, but both have been converted to adaptive landscapes by adding the dimension of light-interception efficiency to the two morphological dimensions.

Simpson (1944, 1953) made the conceptual jump of modelling macro-evolutionary phenomena in geological time on adaptive landscapes, as discussed in Chapter 3. In the next chapter we shall actually analyse such phenomena through the usage of theoretical morphospaces.

6

Analysing evolutionary phenomena in theoretical morphospaces

> There is no necessary link between theoretical morphology and adaptation ... Theoretical morphology is about form and possibility; adaptation is about function and efficiency in the realm of the actual. Adaptation can certainly shape the occupation of morphospace; but because the model underlying a morphospace can produce nonfunctional possibilities, the morphogenetic process itself is nonadaptive ... the [model] parameters themselves cannot be adaptive, only their values can.
>
> *Eble (2000, p. 524)*

Analysing evolution in geological time

In Chapter 3 we saw that we could use the adaptive landscape concept to conceptually model large scale evolutionary phenomena, macroevolutionary phenomena that may involve thousands of species or operate across millions of years of geological time. In this chapter we shall see that the analytical techniques of theoretical morphology allow us to actually analyse the adaptive significance of macroevolutionary phenomena, and not simply to conceptually model those phenomena.

Analysing convergent evolution in theoretical morphospace

The geometry of the helix is ubiquitous in nature – an incredible number of biological structures on all scales, from molecules to entire animals, have evolved helical structures (just a few years ago, in 2003, we observed the 50th anniversary of the discovery that the coding mechanism of life itself, DNA, has a helical structure). Within the Bryozoa, a group of colonial marine animals, helical colonies have convergently, independently, evolved in no less than six separate genera in distantly related higher taxa, scattered across a span of time comprising some 400 million years (Fig. 6.1).

McKinney and Raup (1982) created a mathematical model and computer program for helical-colony form simulation, which they then used to create a helical-byrozoan theoretical morphospace. Figure 6.2 illustrates two dimensions of the morphospace, and the spectrum of hypothetical helical-colony forms that can be produced by varying the model parameter values *ELEV*, the rate of climb of the helix, and *BWANG*, the angle between the central helical axis and the filtration-sheet whorls of the colony. Note that as the value of *ELEV* becomes larger, the helical shape of the colony becomes more elongate and the colony's filtrations-sheet whorls more stretched-out and less overlapping. As the value of *BWANG* becomes larger, the filtration-sheet whorls of the colony are oriented at higher and higher angles to the central helix, until they eventually project out at right angles from the helical axis of the colony.

In Figure 6.3 are given the boundary polygons of measurement data taken from seven different groups of actual helical bryozoans, both extinct and alive, within the morphospace. Note the overlap-region of the polygons in the centre of the figure: these bryozoans have not only convergently evolved helical colonies, they have *repeatedly evolved helical colonies that have the same geometry*, over and over again. The computer simulation given in the upper right of the figure illustrates this convergently evolved geometry.

Now note the four computer simulations given in Figure 6.4. These simulations represent *nonexistent* colony morphologies; these four regions of the morphospace are empty of bryozoans. Thus the theoretical morphospace can show us not only what organic form nature has produced over and over again, it can also reveal to us biological form that is theoretically possible, but never produced by nature. Analysis of these nonexistent colony morphologies reveals that they represent nonfunctional geometries for the filter-feeding mode of life of marine bryozoans (McKinney and McGhee, 2003). Colony geometries with very low values of *BWANG* and *ELEV* have filtration-sheet whorls that are nested one within another, and that extensively overlap one another (see Fig. 6.2). These geometries inhibit water flow through the colony, and produce dead-water zones in the colony centre. Colony geometries with very high

Figure 6.1. Illustration of fossil and living bryozoans that have convergently evolved helical colony forms. Figure 6.1A illustrates the central helical screw of a Carboniferous species of *Archimedes*, where the delicate fronds of the filtration-sheet whorls have been broken away. Figures 6.1B and 6.1C shows two colonies in which the filtration-sheet whorls of *Archimedes* colonies have been preserved, compressed into the sediment in Fig. 6.1B and preserved within the infilling sediment in Fig. 6.1C. Figure 6.1D illustrates the central helical screw of the Eocene species *Crisidmonea archimediformis*. Figures 6.1E and 6.1F illustrate two living genera of bryozoans with species having helical colony geometries, the Australian *Retiflustra cornea* in Fig. 6.1E and the North American *Bugula turrita* in Fig. 6.1F. Length of scale bars is 10 millimetres in Figs. 6.1A–6.1E, and one millimetre in Fig. 6.1F.
Source: From McKinney and McGhee (2003).

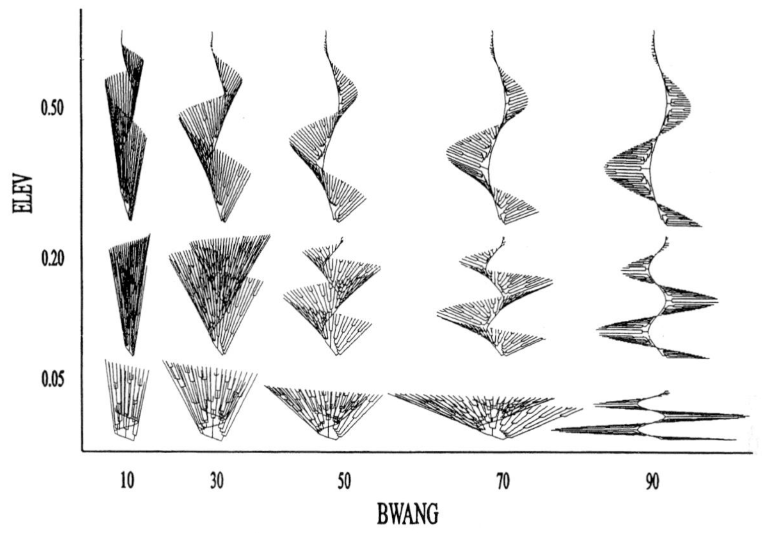

Figure 6.2. A theoretical morphospace of hypothetical helical colony form in bryozoans. The two morphological trait dimensions of the morphospace are *ELEV*, the rate of climb of the helix, and *BWANG*, the angle between the central helical axis and the filtration-sheet whorls of the colony. For purposes of illustration, the dimensional axes are not arithmetically scaled. Source: From Raup, McGhee and McKinney (2006), where the code for producing these simulations may be found.

values of *BWANG* have very poor filtration potential, as most water flows through the colony without ever crossing the filtration sheet whorls. Last, colony geometries with very high values of *ELEV* have the same poor filtration potential, in that the filtration-sheet whorls are too stretched-out (see Fig. 6.2), and much water passes through the colony without being filtered for food.

In contrast, the convergently re-evolved colony form shown in Figure 6.3 has the maximum efficiency in filtration potential – its filtration-sheet whorls are oriented at an angle that best intersects all water flow through the colony, without at the same time impeding water flow. Its helical repeat-distance, a function of its *ELEV* value, best spaces the filtration-sheet whorls out for maximum filtration surface-areas with minimum areal overlap.

The analytical techniques of theoretical morphology allow us to take the heuristic concept of convergent evolution on an adaptive landscape (Fig. 3.1) and to apply it to the analysis of the evolution

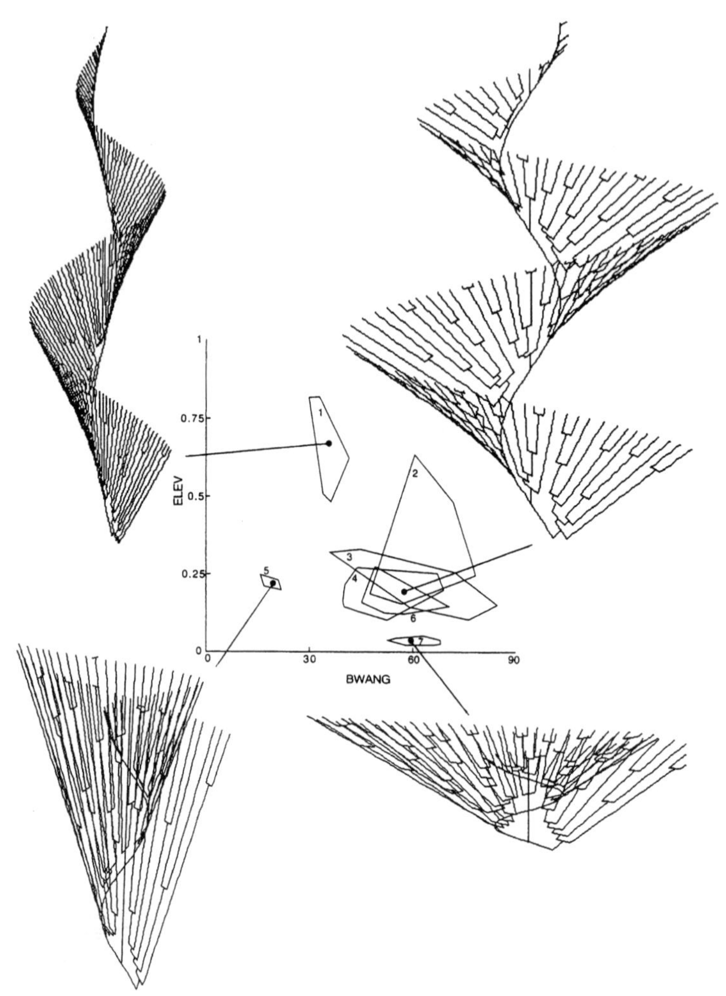

Figure 6.3. Boundary data polygons of data taken from 208 helical colony forms that have been evolved in seven different groups of marine bryozoans. Computer simulations of existent colony form within the morphospace are also illustrated: the simulation in the upper right illustrates the morphology most frequently attained by convergent evolution, shown by the overlapping boundary polygons of morphologies evolved in four separate groups of bryozoans. The two simulations on the left and one simulation in the lower right illustrate more rarely evolved bryozoan morphologies (each present in only one species).
Source: From Raup, McGhee and McKinney (2006).

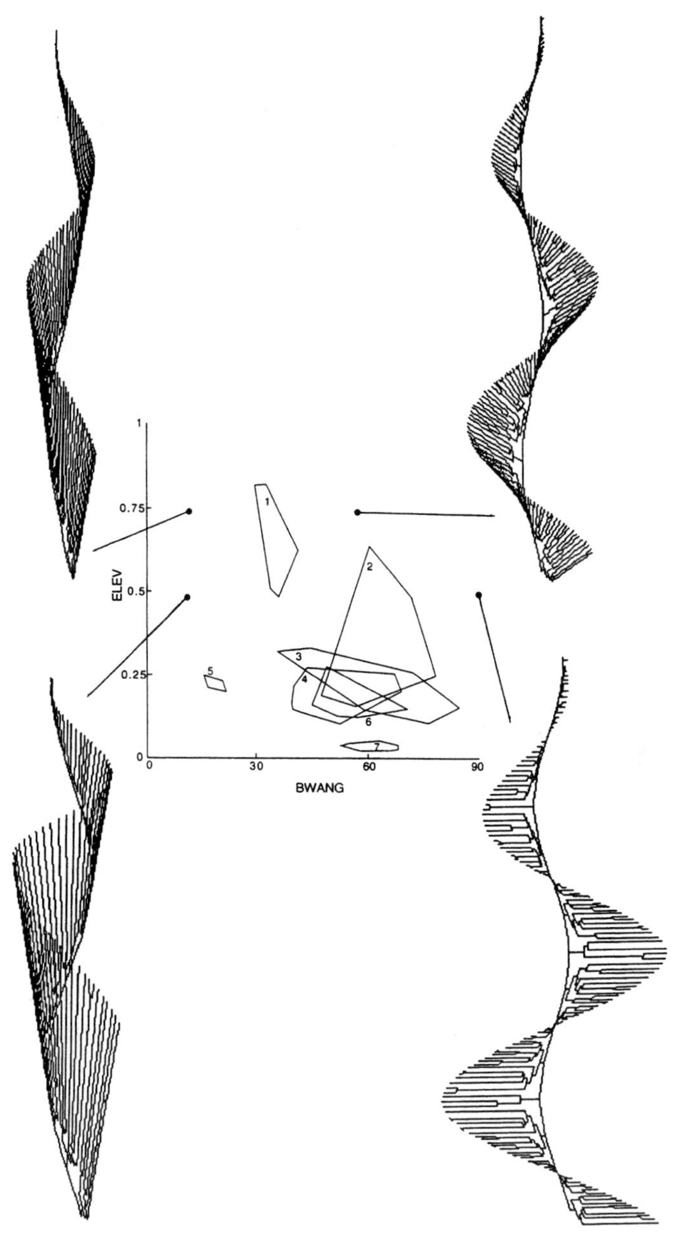

Figure 6.4. Computer simulation of nonexistent helical colony forms within the theoretical morphospace. Although these morphologies are geometrically possible, they have never evolved as organic forms within the bryozoans.
Source: From Raup, McGhee and McKinney (2006).

of life (Figs. 6.3 and 6.4). In essence, the overlapping boundary polygons of helical-colony morphologies evolved within the *Bryozoa* in the past 400 million years, illustrated in Figure 6.3, are the apex and upper slope regions of an adaptive peak of helical-colony form. The three colony morphologies shown on the left margin and lower right in Figure 6.3 represent the lower slope regions of the adaptive peak – they function, but not as well as the peak morphology, and each is only found in one species of bryozoan, respectively. And last, the four computer simulations given in Figure 6.4 show us the colony geometries that lie out on the flat plane of the adaptive landscape, the region of nonfunctional helical colony forms. These actual convergent–evolution relationships are schematically summarized in Figure 6.5.

Analysing iterative evolution in theoretical morphospace

In Chapter 3 we considered the phenomenon of iterative evolution, in which a group of daughter species with very similar morphologies repeatedly originate, one after another, from an ancestral species that in itself may change very little in geological time. In iterative evolution a similar evolutionary pathway repeated appears in an adaptive landscape, from one region to another region, rather than the evolutionary convergence on a similar morphology from multiple, and different, regions within the adaptive landscape (Fig. 3.1).

Iterative evolution was particularly common in the extinct ammonites. In Figure 6.6 are illustrated three separate groups of Jurassic ammonites (the leioceratines, graphoceratines and sonniniids) that repeatedly evolved descendant species with smooth shells, having narrow umbilici and high whorl expansion rates, from ancestral species that had ornamented shells with wide umbilici and low whorl expansion rates. The ancestral species also all had shells with wide, oval apertures, whereas the descendant species all evolved shells with compressed, narrow apertures.

In Figure 6.7, the vectors represent morphological data taken from four groups of Jurassic ammonites (the hammatoceratids have been added to the three groups shown in Fig. 6.6), plotted in a theoretical morphospace of ammonoid form. The dimensions of this morphospace are different from the *W-D* dimensions that we have used previously in our analyses of the evolution of ammonite form (Fig. 5.1), in that the parameter *W* has been replaced by *S*, the shape

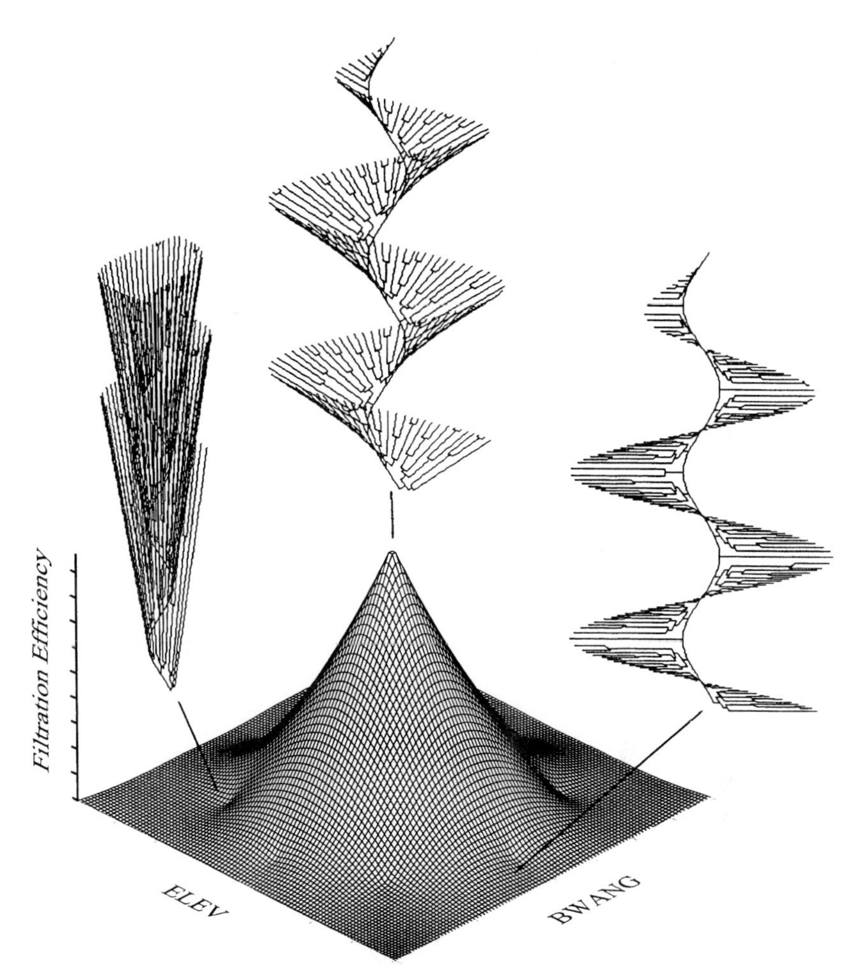

Figure 6.5. A schematic summary of the convergent evolution of helical-colony forms within the Bryozoa. Six different genera of actual bryozoans, starting from different morphological regions of the morphospace, have convergently climbed the adaptive peak of maximum filtration efficiency within the morphospace, and convergently re-evolved the colony geometry illustrated in the top centre of the figure. The hypothetical colony forms shown to the right and left of the centre simulation have very poor filtration potentials, and have never evolved in real bryozoans.

of the aperture of the shell. Raup (1967) modelled the aperture of ammonoid shells as ovoid to circular in shape, where S is simply defined as the ratio of the minimum axis of the aperture to the maximum axis. In a hypothetical shell having a circular aperture, the minimum and

Figure 6.6. Iterative evolution of smooth-shelled ammonite species with narrow umbilici and high whorl expansion rates (right side of figure) from ornamented-shelled species with wide umbilici and low whorl expansion rates (left side of figure) in Jurassic ammonites. Three separate iterative-evolutionary sequences are illustrated: ammonites in the top row are leioceratines, in the middle row are graphoceratines, and in the lower row are sonniniids. Modified from "Iterative evolution of Middle Jurassic ammonite fauna", by U. Bayer and G. R. McGhee, in *Lethaia*, www.tandf.no/leth, 1984, volume 17, pp. 1–16. Copyright © 1984 by Taylor & Francis AS and reprinted with the permission of the publisher.

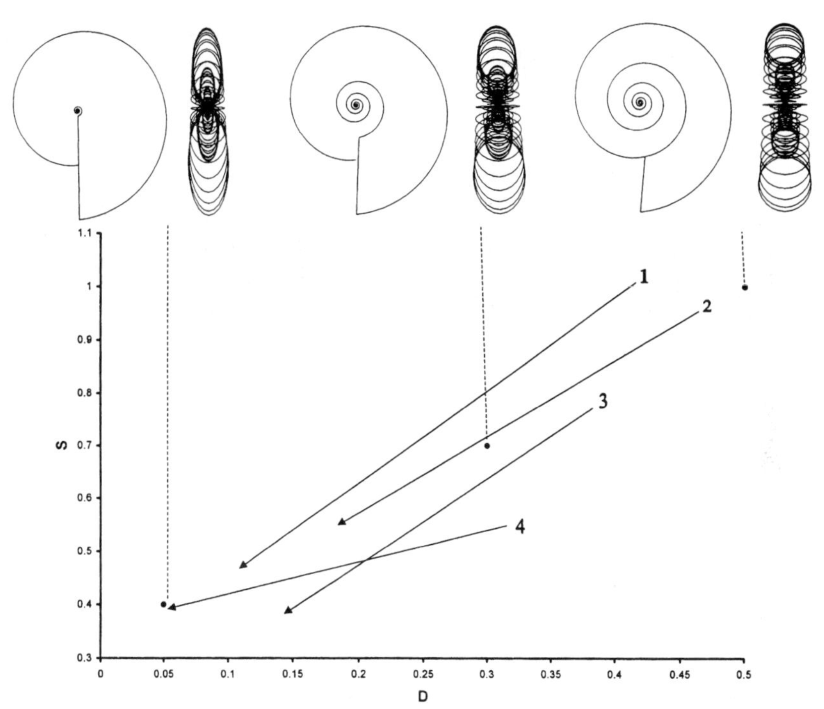

Figure 6.7. Iterative morphological evolution in four separate groups of Jurassic ammonites in theoretical morphospace. The morphological trait dimensions of the morphospace are *S*, the shape of the shell aperture, and *D*, the distance from the coiling axis to the aperture. The vectors within the morphospace give the morphological range and evolutionary direction of change in morphologies found in (1) sonniniid species, (2) hammatoceratid species, (3) graphoceratine species, and (4) leioceratine species. All four groups of ammonites iteratively moved from the high-*S*, high-*D* region of the morphospace to the low-*S*, low-*D* region of the morphospace, as illustrated by the vectors. Computer simulations illustrate the morphologies present in select regions of the morphospace. Data from Bayer and McGhee (1984).

maximum axes have the same magnitude, and the value of *S* is thus 1.0. Computer simulations of different shell forms in different *S* and *D* regions of the morphospace are illustrated in Figure 6.7.

With each of the four data vectors shown, ammonites repeatedly evolved descendant species in the low-*S*, low-*D* region of the morphospace from ancestral shell morphologies in the high-*S*, high-*D* region of the morphospace. This iterative pattern of evolution produces the four strikingly parallel evolutionary pathways seen within the morphospace. The evolutionary pathways are not identical – the sonniniid and hammatoceratid ammonite species started out in a higher-value *S* and *D* region of the morphospace than the graphoceratines and leioceratines, and they did not evolve shells with as low values of *S* that the graphoceratines and leioceratines did – yet the parallelism is still uncanny.

What evolutionary mechanism could produce such a phenomenon? In this case the actual analysis of iterative evolution of ammonite morphologies in theoretical morphospace (Fig. 6.7) preceded the construction of the adaptive-landscape conceptual model to explain the phenomenon (Fig. 3.3), not vice versa! It turns out that the ammonites in the high-*S*, high-*D* region of the morphospace are deep-water, open-oceanic species, and the ammonites in the low-*S*, low-*D* region of the morphospace are shallow-water, continental-shelf species. Each of the iterative cycles of morphological evolution (Fig. 6.7) were triggered by repetitive transgressive-regressive cycles of sea level (Fig. 3.3), in which shallow-water shelf habitats repeatedly appeared and disappeared. Each time the shallow-water shelf habitat appeared, a deep-water species evolved a daughter species that invaded it (Bayer and McGhee, 1984, 1985; McGhee, Bayer and Seilacher, 1991).

Analysing biodiversity crises in theoretical morphospace

Catastrophic losses of biodiversity have occurred many times in the history of life on Earth. The five most severe losses of biodiversity in geological time are termed the 'Big Five': these are the End-Ordovician, the Late Devonian (between the Frasnian and Famennian Stages within the Late Devonian), the End-Permian (which ended the Palaeozoic Era), the End-Triassic, and the End-Cretaceous biodiversity crises (which ended the Mesozoic Era). Three of the Big Five biodiversity crises were trigged by dramatic jumps in the extinction rate of species in geological time: the End-Ordovician, the End-Permian and the End-Cretaceous, which are also known as periods of 'mass extinction' (Bambach, Knoll and Wang, 2004). In contrast, the Late Devonian and End-Triassic biodiversity crises were ecologically more complex, being triggered by a precipitous decline in speciation rates operating in concert with elevated extinction rates (McGhee, 1988; Bambach, Knoll and Wang, 2004). With the notable exception of the End-Permian, the magnitude of biological diversity loss in each crisis has declined with geological time, indicating a biosphere that is becoming more extinction resistant. Quite to the contrary, the ecological disruption produced by each crisis has increased with geological time, indicating a global ecosystem that is becoming increasingly more interdependent and integrated (McGhee *et al.*, 2004). Outside the Big Five, other notable biodiversity crises occurred in the Late Cambrian, End-Devonian, Mid-Carboniferous, Early Eocene and the Pleistocene. Indeed, some would add the current global loss of biodiversity, produced by human destruction of habitat space, to the list as well.

Some of the most spectacular 'boom and bust' cycles of biodiversity increase and decrease in geological time are exhibited by the ammonoids. These animals barely survived the Late Devonian, End-Devonian, End-Permian biodiversity crises, but each time they re-evolved new species following the crisis and eventually recovered comparable biodiversities. However, they did not survive the End-Cretaceous biodiversity crisis. The only even remotely similar cephalopods with external shells that exist today are the species of the single genus *Nautilus*, a close relative of the extinct ammonoids.

Analyses of the morphological effects of the Late Devonian, End-Devonian and End-Permian biodiversity crises on ammonoid evolution in theoretical morphospace have been conducted by Korn (2000), Saunders, Work and Nikolaeva (2004) and McGowan (2004). In both the Late Devonian and End-Devonian biodiversity crises, the few genera that survived had geometries with very low values of D (Fig. 6.8), located near the hydrodynamic-efficiency adaptive peak in the left region of the morphospace (Fig. 5.2). Shells in this region of the morphospace are involute, with small umbilici (simulation given in Fig. 6.8). From such a slim survival margin, the ammonoids explosively evolved new species post-crisis such that the diversity of shell geometries seen in the later Famennian and the later Tournaisian (Early Carboniferous) ammonoids is similar to that seen in the morphospace prior to the biodiversity crisis (Saunders, Work and Nikolaeva, 2004). In particular, the empty adaptive peak on the right in the morphospace (Fig. 6.8) is quickly reinvaded.

In contrast, the two genera of ammonoids that survived the End-Permian biodiversity crisis were located near both of the maximum swimming-efficiency adaptive peaks in the morphospace (cf. Figs. 6.9 and 5.2). Interestingly, the ammonoid genus near the swimming-efficiency adaptive peak on the left, the polecanitid

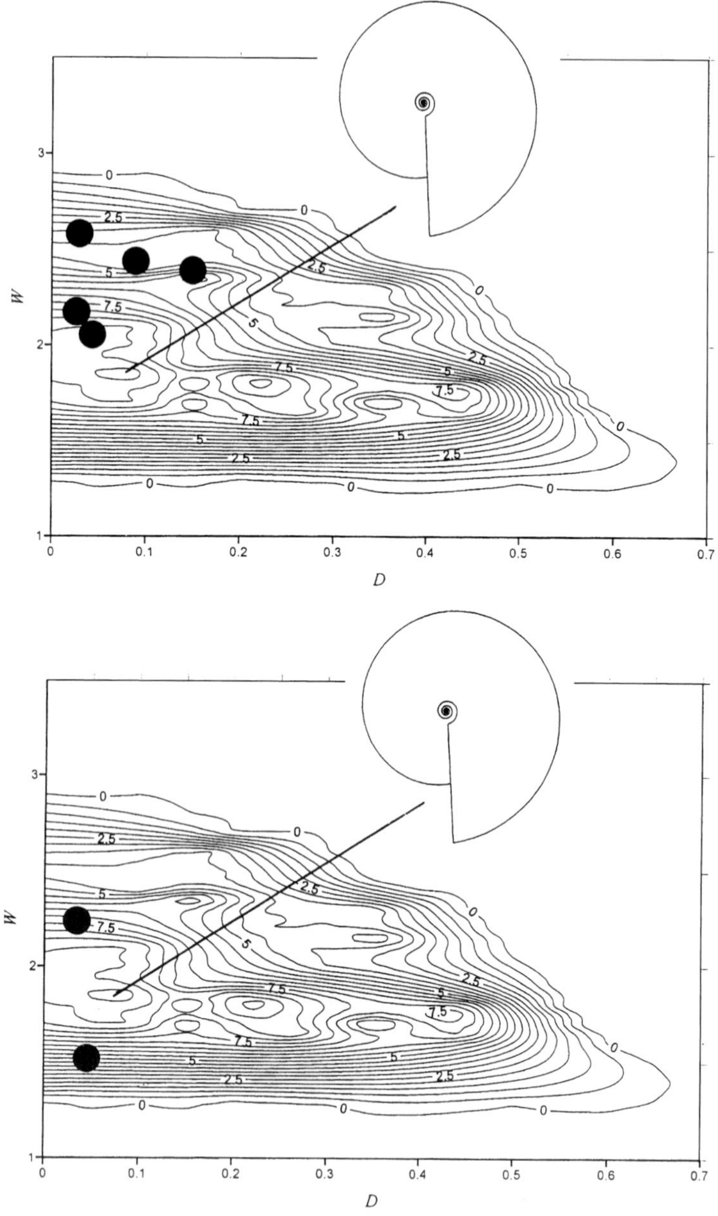

Figure 6.8. Morphospace position of the five ammonoid genera that survived the Late Devonian biodiversity crisis (black points plotted in the upper figure) and the two genera that survived the End-Devonian biodiversity crisis (black points plotted in the lower figure) shown with respect to the total morphological frequency distribution seen in Palaeozoic ammonoids. All survivors are close to only one of the swimming-efficiency peaks shown in Figure 5.2. Data from Saunders, Work and Nikolaeva (2004).

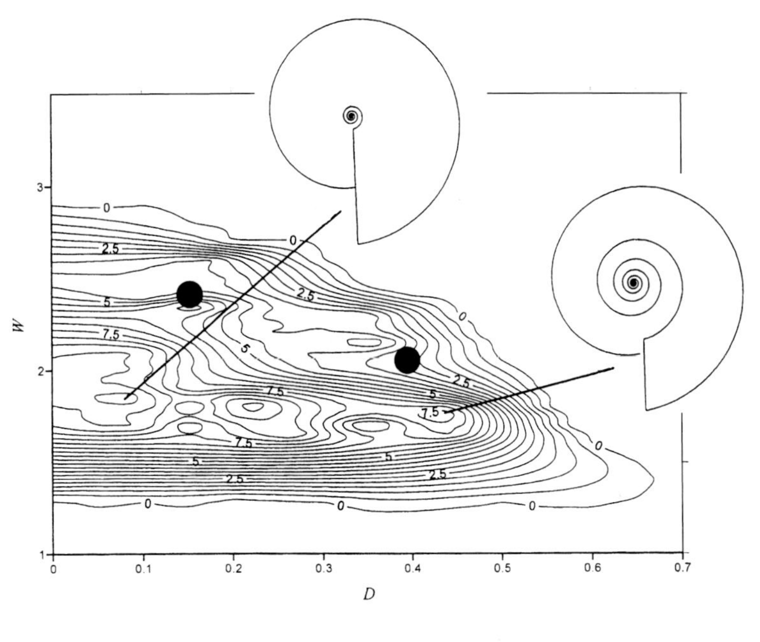

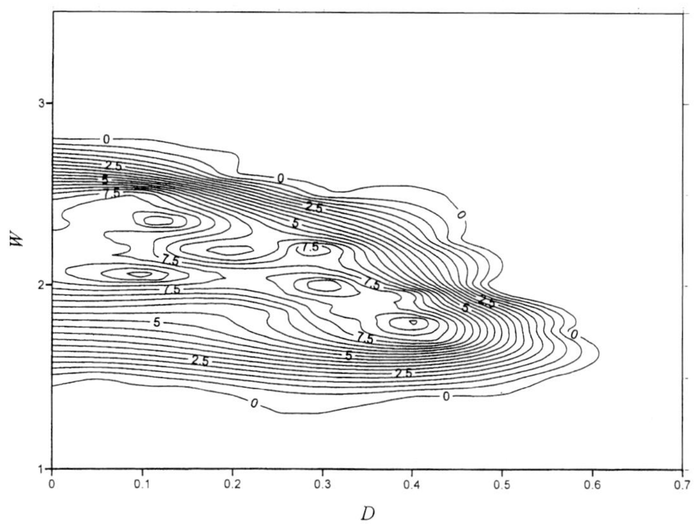

Figure 6.9. Morphospace position of the two ammonoid genera that survived the End-Permian biodiversity crisis (black points plotted in the upper figure) shown with respect to the total morphological frequency distribution seen in Palaeozoic ammonoids (data from Saunders, Work and Nikolaeva, 2004). The lower figure shows, for comparison, the morphological frequency distribution eventually re-evolved by post-crisis Triassic ammonoids. Both of the two swimming-efficiency peaks shown in Figure 5.2 are reoccupied by ammonoids in the Triassic.
Source: Data from McGowan (2004).

Episageceras, did not survive long into the Triassic, and left no descendants. All of the subsequently evolved Mesozoic ammonoids were the descendants of the ammonoid that survived near the swimming-efficiency adaptive peak in the right region of the morphospace – the ceratitid *Xenodiscus* (Saunders, Work and Nikolaeva, 2004). Shells in this region of the morphospace have moderate whorl overlap and wide umbilici (simulation given in Fig. 6.9). Yet even so, subsequent new species evolution in the later Triassic produced shell geometries that reinvaded the adaptive peak on the left, such that the frequency distribution of morphologies found in later Triassic ammonoids is similar to that found in the Palaeozoic ammonoids (Fig. 6.9).

It is not clear why ammonoids with shell geometries near the swimming-efficiency adaptive peak in the right region of the morphospace did not survive the Late Devonian and End-Devonion biodiversity crises (Fig. 6.8), or why the surviving genus near the adaptive peak in the left region of the morphospace following the End-Permian biodiversity crisis (Fig. 6.9) died out, leaving no descendants. These survival patterns may simply reflect the role of chance, or *historical contingency* (Gould, 1989), in ammonoid morphological evolution.

Of major importance, however, is the fact that following each of the three biodiversity crises the few slim survivors subsequently repopulated the empty regions of the morphospace. Two critical points are revealed in the observed pattern of morphological evolution of the ammonoids during periods of biodiversity crisis: one, the *stability* of the adaptive landscape upon which the ammonoids evolved, and two, the fact that ammonoids from very different phylogenetic lineages *convergently re-evolved* the same shell geometries following the biodiversity crisis. The two hydrodynamic-efficiency adaptive peaks in ammonoid theoretical morphospace (Fig. 5.2) have thus continued to exist for the 145 million year history of Palaeozoic ammonoid evolution. Each time one of the peaks was vacated in a biodiversity crisis, the surviving ammonoids subsequently reinvaded the empty peak post-crisis, demonstrating the continued stability of the peak. Thus the analysis of the actual pattern of morphological evolution exhibited by the ammonoids in theoretical morphospace contradicts the expectations of the Red Queen model of evolution on adaptive landscapes, with its constantly moving adaptive peaks (as discussed in Chapter 2). Likewise, the pattern of ammonoid evolution contradicts the Strathmann (1978) model whereby adaptive peaks that are vacated during a biodiversity crisis remain empty following the crisis because organisms in adjacent regions of the adaptive landscape have become too specialized to reinvade the empty adaptive peak (as discussed in Chapter 3).

The fact that ammonoids from very different phylogenetic lineages convergently re-evolved the same shell geometries following each biodiversity crisis argues against any strong effect of *phylogenetic constraint*, at least in ammonoid evolution. We shall examine the concept of phylogenetic constraint in more detail in Chapter 7. Suffice it here simply to note that the highly involute-shelled ammonoid survivors of the Order Goniatitida convergently re-evolved moderately involute descendants following the Late Devonian biodiversity crisis (the Order *Clymeniida*) and the End-Devonian biodiversity crisis (the Order *Prolecanitida*), whereas the moderately involute-shelled survivors of the Order *Ceratitida* convergently re-evolved highly involute-shelled descendants, similar in geometry to the extinct Order *Goniatitida*, following the End-Permian biodiversity crisis (Saunders, Work and Nikolaeva, 2004; McGowan, 2004).

Analyses of the morphological effects of biodiversity crises in theoretical morphospace have also been conducted with brachiopods (McGhee, 1995). Interestingly, and analogous to the ammonoids, the morphological response of brachiopods to the End-Permian and End-Cretaceous biodiversity crises was very similar, even though very different phylogenetic lineages are involved in these two crises, and even though the crises are separated by some 180 million years of geological time. In both cases a major loss of morphological diversity occurs, though not near as severe as that seen in the ammonoids (Figs. 6.8 and 6.9). The area extent of the morphological frequency distribution of brachiopod shell geometries within the theoretical morphospace shrank by 8.7% from the Permian to the Triassic, and by 10.2% from the Cretaceous to the Palaeogene (McGhee, 1995). Moreover, the boundaries of the morphological frequency distributions shift within the morphospace, and the direction of that shift was the same for both biodiversity crises.

The outer boundary of the morphological frequency distributions of brachiopod shell geometries before and after the End-Permian biodiversity crisis are given in Figure 6.10. A major loss of brachiopods having highly inequivalved shells with very flat dorsal valves (high dorsal W values) occurred from the Permian to Triassic (Fig. 6.10). On the other hand, new brachiopod morphologies appeared

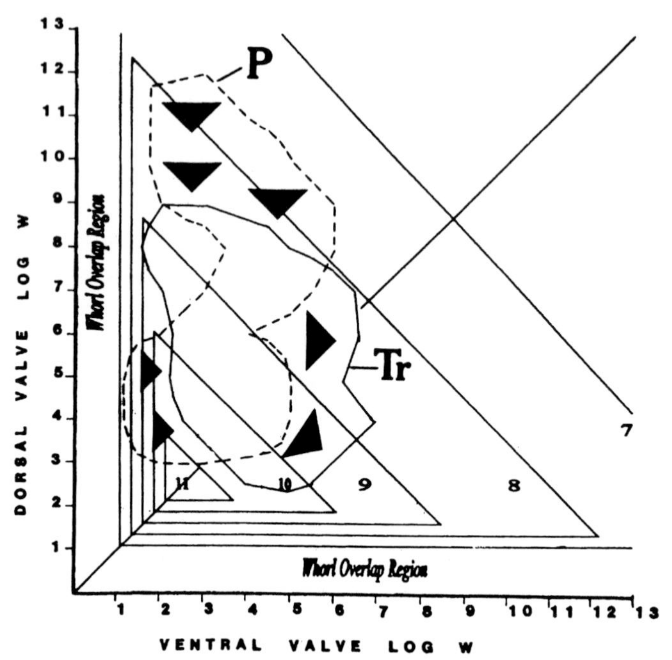

Figure 6.10. Shifting morphological frequency distribution. The Permian boundary of the morphological frequency distribution of brachiopod shell geometries is given by the dashed line, and the Triassic boundary is given by the solid line, relative to the contours of shell volume-to-surface-area ratios for differing geometries within the morphospace (Fig. 5.4). Permian brachiopods evolved shells far away from the adaptive peak of maximum volume-to-surface-area ratios, whereas Triassic brachiopods retreated from the morphospace region having shells with larger surface areas and smaller volumes (arrows).
Source: Data from McGhee (1995).

in the Triassic with flatter ventral valves (higher ventral W values), such that the median ventral valve position shifted from a value of the logarithm of W of 3.16 to a value of 4.34, which is one of the few statistically significant shifts (at the 1% level) in median brachiopod shell form which occurs in geological time (McGhee, 1995). The end result of the biodiversity crisis was that brachiopods made a major retreat from dorsal high W regions of the morphospace but spread into ventral higher W regions. That is, the entire frequency distribution of brachiopod shell forms shifted down and to the right within the morphospace, toward the region of equiconvexity, where shell forms with ventral and dorsal valves of equal W values are located. Post-extinction Triassic brachiopod shells were more equibiconvex, and more centred on the slopes of the adaptive peak of maximum volume-to-surface-area ratios within the morphospace (Fig. 6.10). A similar shift within the morphospace occurs during the End-Cretaceous biodiveristy crisis (McGhee, 1995).

From a functional perspective it appears that brachiopods in their evolutionary history have repeatedly evolved less-than-optimum shell forms (as discussed in Chapter 5) in the periods of geological time between biodiversity crises. That is, periods of brachiopod morphological diversification were characterized by expansion of the boundaries of their morphological frequency distribution into regions of the theoretical morphospace with shell forms having lower volume-to-surface-area ratios (Fig. 6.10).

The biodiversity crises at the end of the Palaeozoic and Mesozoic are evidenced by the retreat and contraction of the boundaries of the morphological frequency distribution back to regions of the theoretical morphospace characterized by shell forms with more equibiconvex, spherical morphologies. Rather than vacating and reinvading a stable adaptive peak, the evolutionary pattern seen in biodiversity crises in the ammonoids (Figs. 6.8 and 6.9), the brachiopods diversified out into their morphospace during good times, only to retreat back up the slopes of their adaptive peak during times of crisis (Fig. 6.10).

Are there limits to the amount of movement a group of organisms might potentially make in theoretical morphospace, other than those of function? This is a question we shall take up in the next chapter, where we shall examine the concept of *evolutionary constraint*.

7

Evolutionary constraint in theoretical morphospace

> Our results strongly support the hypothesis that the essential elements of organic structure are highly constrained by geometric rules, growth processes, and the properties of materials. This suggests that, given enough time and an extremely large number of evolutionary experiments, the discovery by organisms of 'good' designs – those that are viable and that can be constructed with available materials – was inevitable and in principle predictable.
>
> *Thomas and Reif (1993, p. 342)*

Potential causes of empty morphospace

One surprising revelation of many actual theoretical morphospace analyses is the degree to which *the morphospace is empty*. Empty morphospace is not simply a conceptual model, a heuristic construct in visualizing the potential evolution of life. It is an empirical reality, and part of the power of theoretical morphospace analysis is the ability of this type of analysis to reveal empty morphospace to us (McGhee, 1999, 2001a). The subsequent analysis of empty morphospace involves the concept of *evolutionary constraint*, the concept that there are limits to the spectrum of possible evolutionary change. We shall examine the potential causes of evolutionary constraint in this chapter.

Some confusion exists at present concerning the concept of evolutionary constraint, and the various types of constraint (see the reviews of Maynard Smith *et al.*, 1985; Antonovics and van Tienderen, 1991; McKitrick, 1993; Schwenk, 1995; Blomberg and Garland, 2002; Cubo, 2004). One group of potential constraints has been variously called developmental, architectural, fabricational, constructional, ontogenetic and morphogenetic. These have been variously contrasted with or equated to phylogenetic constraints historical constraints, canalization constraints and phylogenetic inertia. The resulting confusion has prompted one set of reviewers to tongue-in-cheek propose 'onto-ecogenophyloconstraints' to cover all concepts (Antonovics and van Tienderen, 1991).

Even where the concept of constraint in evolution is accepted, confusion often still exists as to the actual type of constraint in action. For example, Ciampaglio (2002, p. 182) describes the fact that 'bivalves must have nonoverlapping whorls in order to achieve a functioning hinge' (a fact that we considered in some detail in Chapter 5 with respect to another group of bivalved organisms, the brachiopods) as an example of developmental constraint; that is, animals with articulated bivalved shells are constrained to occur only in the morphospace characterized by shells with morphologies with whorls that do not overlap. It is indeed true that bivalve molluscs, for example, are unable to develop articulated shells with valves that have overlapping whorls, but not for the reason that bivalve molluscs do not possess the genetic coding needed to produce valves with overlapping whorls (phylogenetic constraint) or that the morphogenesis of such a valve geometry is not possible (developmental constraint). Instead, bivalve molluscs and brachiopods cannot develop shells with valves with overlapping whorls *because it is geometrically impossible to articulate those two valves into a bivalved shell* (geometric constraint).

In this chapter I shall argue that four conceptually distinct types of evolutionary constraint exist: geometric, functional, phylogenetic and developmental. The first two constraints are *extrinsic* functions of the laws of physics and geometry, whereas the latter two constraints are *intrinsic* functions of the biology of specific organisms.

Modelling geometric constraint

In this chapter we shall see also that the analytical techniques of theoretical morphology allow us to take a spatial approach to the concept of evolutionary constraint. To explicitly define the concepts of evolutionary constraint considered here, I shall use Venn diagrams and set theory. A given biological form, f, may be described by a set of measurements taken from that form. Each type of measurement, such as length, width, or height, can be considered as a form dimension.

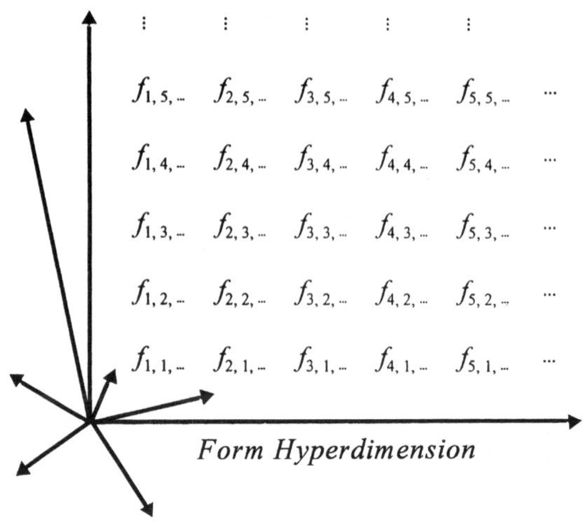

Figure 7.1. A theoretical hyperdimensional space of possible form
Each dimension of the space represents a morphological trait that may be measured on a given biological form, f. All possible coordinate combinations within the hyperdimensional space represent the set of all possible biological forms. Although only eight dimensions are shown in this schematic diagram, the dimensionality of an actual hyperspace of form will be much larger.

The total set of the possible dimensions of form can be used to construct a hyperdimensional space of possible form (Fig. 7.1). All coordinate combinations within this space represent a universal set of form, U. Some of the coordinate combinations within the total hyperspace of form represent the set of geometrically possible forms: $GPF = \{f | f = \text{geometrically possible forms}\}$ (Fig. 7.2). Other coordinate combinations within the total hyperspace represent the set of geometrically impossible forms: $GIF = \{f | f = \text{geometrically impossible forms}\}$.

The regions of impossible and possible form within the total hyperspace of form do not overlap one another; that is, a given set of coordinates within the hyperspace cannot simultaneously represent a possible geometry and an impossible geometry (Fig. 7.2). The sets of form GIF and GPF are thus compliments of each other:

$$GIF \cup GPF = \{f | f \in GIF \quad \text{or} \quad f \in GPF\} = U,$$
$$GIF \cap GPF = \{f | f \in GIF \quad \text{and} \quad f \in GPF\} = \oslash.$$

The boundary between the sets GIF and GPF is here designated as the *geometric constraint boundary* (Fig. 7.2).

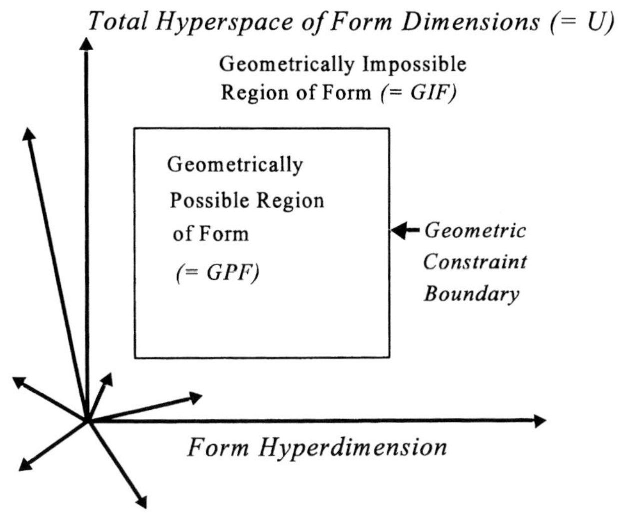

Figure 7.2. Geometric constraints. Geometric constraint as the boundary between the set of geometrically impossible forms (*GIF*) and the set of geometrically possible forms (*GPF*) in the total hyperspace of form dimensions (the universal set of form, *U*).
Source: From McGhee (2006).

Modelling functional constraint

Within the region of geometrically possible forms there exists two subregions: a region of forms that are functional in nature, and a region of nonfunctional forms (Fig. 7.3), and the totality of these forms represent the two sets $FPF = \{f \mid f =$ functional possible forms$\}$ and $NPF = \{f \mid f =$ nonfunctional possible forms$\}$. The word ‘nonfunctional’ is used here in the sense of lethal function; that is, possession of nonfunctional form does not allow the organism to survive in nature.

The regions of functional and nonfunctional form within the region of geometrically possible form do not overlap one another; that is, a given set of coordinates within the geometrically possible region cannot simultaneously represent a form that is both functional and nonfunctional. Thus the sets of form *FPF* and *NPF* are compliments of each other, and contained within *GPF*:

$$FPF \cup NPF = \{f \mid f \in FPF \quad \text{or} \quad f \in NPF\} = GPF,$$
$$FPF \cap NPF = \{f \mid f \in FPF \quad \text{and} \quad f \in NPF\} = \oslash.$$

The boundary between the sets *NPF* and *FPF* is here designated as the *functional constraint boundary* (Fig. 7.3).

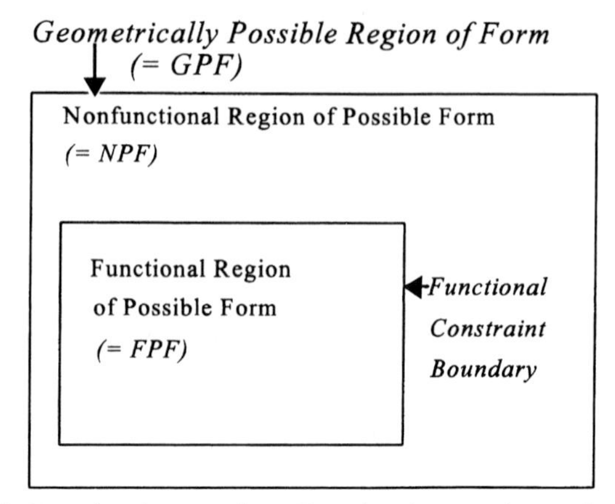

Figure 7.3. Functional constraints. Functional constraint as the boundary between the set of nonfunctional possible form (*NPF*) and the set of functional possible form (*FPF*) within the set of geometrically possible form (*GPF*).
Source: From McGhee (2006).

Modelling phylogenetic constraint

The concepts of geometric constraint and functional constraint are here considered to belong to the class of *extrinsic constraint*, where extrinsic constraints are those imposed by the laws of physics and geometry. Extrinsic constraints exist whether any actual biological form encounters them or not. A separate class of constraint is that of *intrinsic constraint*, where intrinsic constraints are those imposed by the biology of a specific organism. Instrinsic constraints do not exist in the absence of an actual organism.

At least two conceptually different types of intrinsic constraint exist: phylogenetic constraint and developmental constraint. In Figure 7.4, *phylogenetic constraint for species x* is illustrated as the boundary of the set of forms that are possible given the genetic coding present in species x, the set $PPFx = \{f|\ f =$ phylogenetically possible forms for species $x\}$. Note that this set has boundaries that hypothetically may cross the boundaries of both functional and nonfunctional possible form (*FPF* and *NPF*), and the boundaries of both geometrically possible and impossible form (*GPF* and *GIF*). Thus, while both sets *FPF* and *NPF* are contained within the set *GPF*, the set *PPFx* is not:

$$PPFx \not\subset GPF, \quad \text{hence} \quad PPFx \not\subset FPF \cup NPF.$$

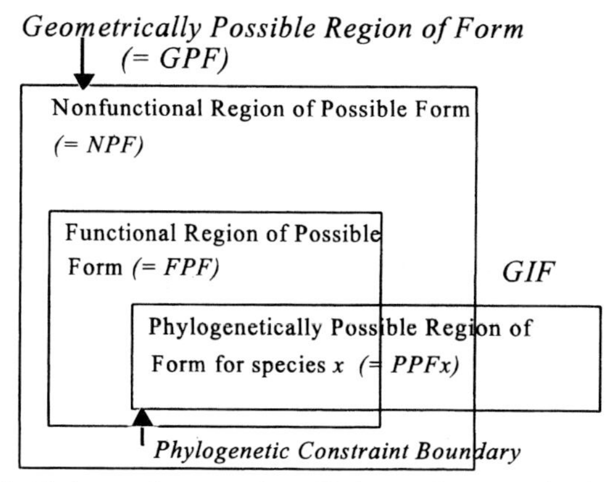

Figure 7.4. Phylogenetic constraints. Phylogenetic constraint for species x as the boundary of the set of forms that are possible given the genetic coding present in species x, ($PPFx$). Note that this set has boundaries that hypothetically may cross the boundaries of both functional and nonfunctional possible form (FPF and NPF), and the boundaries of both geometrically possible and impossible form (GPF and GIF).
Source: From McGhee (2006).

That is, species x may possess the genetic coding to produce functional form, but it may also possess the genetic coding for nonfunctional form. Hypothetically, species x may also possess the genetic coding that could potentially produce coordinate combinations in the total hyperspace of form dimensions that are geometrically impossible (Fig. 7.4).

We can formally consider these three possibilities as set intersections:

$$\begin{aligned} PPFx \cap FPF &= \{f \mid f \in PPFx \quad \text{and} \quad f \in FPF\} \\ &= \text{potential existing form for species } x, \\ PPFx \cap NPF &= \{f \mid f \in PPFx \quad \text{and} \quad f \in NPF\} \\ &= \text{lethal form for species } x, \\ PPFx \cap GIF &= \{f \mid f \in PPFx \quad \text{and} \quad f \in GIF\} \\ &= \text{developmentally impossible form.} \end{aligned}$$

Only the set intersection $PPFx \cap FPF$ represents genetic combinations that will produce functional form in nature. Lethal mutations lie in the set intersection $PPFx \cap NPF$. The hypothetical possible intersection $PPFx \cap GIF$ (i.e. genetic coding for forms that are impossible to develop) is of interest as it leads us to the second intrinsic constraint, that of development.

Modelling developmental constraint

In Figure 7.5, *developmental constraint for species x* is illustrated as the boundary between the two sets $DPFx = \{f| f =$ developmentally possible forms for species $x\}$ and $DIFx = \{f| f =$ developmentally impossible forms for species $x\}$. The regions of developmentally possible and developmentally impossible form within the region of phylogenetically possible form for species x do not overlap one another; that is, a given set of genetic codings cannot simultaneously produce a form that can and cannot be developed. Thus $DPFx$ and $DIFx$ are complements of each other and contained within $PPFx$:

$$DPFx \cup DIFx = \{f| f \in DPFx \text{ or } f \in DIFx\} = PPFx,$$
$$DPFx \cap DIFx = \{f| f \in DPFx \text{ and } f \in DIFx\} = \oslash.$$

Note that although it is hypothetically possible for organisms to possess the genetic coding for forms that are geometrically impossible ($PPFx \cap GIF$), it is developmentally impossible to produce these forms ($DPFx \cap GIF = \oslash$). Thus, unlike $PPFx$,

$$DPFx \subset GPF,$$

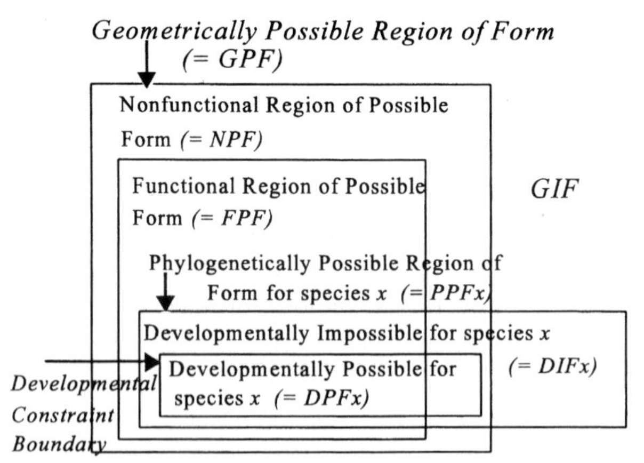

Figure 7.5. Developmental constraints. Developmental constraint for species x as the boundary of the set of forms that are developmentally possible for species x, ($DPFx$), and the set of forms that are developmentally impossible for species x, ($DIFx$), within the set of forms that are phylogenetically possible for species x, ($PPFx$).
Source: From McGhee (2006).

as shown in Figure 7.5. Developmentally possible form must be both phylogenetically possible and geometrically possible, but on the other hand it can either be functional ($DPFx \cap FPF$) or lethal ($DPFx \cap NPF$). That is,

$$DPFx \not\subset FPF.$$

Last, note that the set of developmentally impossible forms for species x, $DIFx$, could hypothetically include forms that geometrically impossible ($DIFx \cap GIF$), forms that are geometrically possible but nonfunctional ($DIFx \cap NPF$), and forms that are both geometrically possible and functional ($DIFx \cap FPF$), but cannot be developed (Figure 7.5).

For the programmer, the distinction between phylogenetic constraint and developmental constraint in morphogenesis may be illustrated by two analogous programming questions. (1) Phylogenetic constraint: is the source code (i.e. the genetic coding) for the morphogenetic program available? (2) Developmental constraint: will the program run, or will it crash? That is, if the code is present we can potentially simulate the form, otherwise we cannot. But, as every programmer knows, possessing the code specifically written to produce a form does not necessarily mean that the program will run on a particular computer, and that the form can be simulated.

The theoretical morphospace characterization of the evolutionary constraint concept developed here is summarized in Table 7.1. The extrinsic constraint boundaries of geometric and functional constraint in form hyperspace are absolute and do not vary in time. That is, the laws of geometry and the physics of swimming are the same today as they were 100 million years ago. This is not true of intrinsic constraint boundaries. Both phylogenetic constraint boundaries and developmental constraint boundaries are intrinsic to the biology of specific organisms and, as organisms evolve with time, their intrinsic constraint boundaries may evolve as well.

Figure 7.6 summarizes the total effect of evolutionary constraint for the development of form by species x within the total hyperspace of form dimensions. The set of potentially existent form for species x, $PEFx$, is thus constrained within the series of set intersections:

$$PEFx = DPFx \cap PPFx \cap FPF \cap GPF = \left\{ \begin{array}{l} f \mid f \in GPF \text{ and } f \in FPF \\ \text{and } f \in PPFx \text{ and } f \in DPFx \end{array} \right\}.$$

Table 7.1. *Theoretical morphospace characterization of the evolutionary constraint concept developed in this chapter*

I. Extrinsic Constraints (imposed by the laws of physics and geometry):
   1. Geometric Constraint: the boundary between form sets *GPF* and *GIF*.
   2. Functional Constraint: the boundary between form sets *FPF* and *NPF*.

II. Intrinsic Constraints (imposed by the biology of specific organisms):
   1. Phylogenetic Constraint for taxon *x*: the boundary between form sets *PPFx* and *PIFx*.
   2. Developmental Constraint for taxon *x*: the boundary between form sets *DPFx* and *DIFx*.

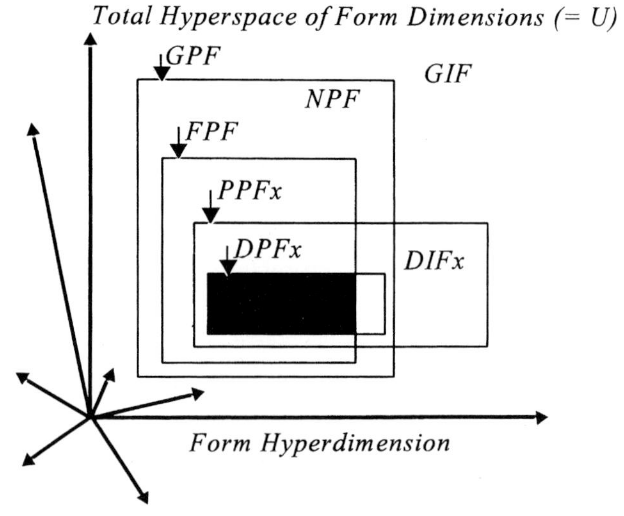

Figure 7.6. Potentially existent form. The set of potentially existent form for species *x*, *PEFx* (black rectangle), constrained as the subset of developmentally possible form for species *x*, (*DPFx*), phylogenetically possible form for species *x*, (*PPFx*), functional possible form (*FPF*) and geometrically possible form (*GPF*) within the total hyperspace of form dimensions (*U*).
Source: From McGhee (2006).

Forms outside the region of the set intersections $DPFx \cap PPFx \cap FPF \cap GPF$ are not possible for species x, but they may be possible for species y, with a different set of phylogenetically possible forms $PPFy$; that is:

$$DPFx \cap PPFx \cap FPF \cap GPF \neq DPFy \cap PPFy \cap FPF \cap GPF.$$

Thus intrinsic constraints will vary from taxon to taxon. Extrinsic constraints do not, however, and all existent form must be in the set $FPF \cap GPF$.

This is not to say that the phenomenon of intrinsic constraint must mean that each separate taxon occupies its own unique region of the total hyperspace of form dimensions. Each taxon has its own unique evolutionary history, thus it is highly unlikely that the set of forms evolved by two taxa will be exactly alike. However, the phenomenon of evolutionary convergence in form, which is extremely common in nature, means that some forms function particularly well in particular environments, and that these forms are repeatedly evolved in many different taxa. That is, although the set of form $DPFx \neq DPFy$, the phenomenon of evolutionary convergence in form means that, for many separate taxa:

$$DPFx \cap DPFy = \{f \mid f \in DPFx \text{ and } f \in DPFy\} \neq \emptyset.$$

This relationship is depicted by the Venn diagram given in Figure 7.7.

Analysing evolutionary constraint in theoretical morphospaces

The analytical techniques of theoretical morphology actually allow us to explore the boundaries of geometric, functional, phylogenetic and developmental constraint within the total hyperspace of form dimensions by the construction of theoretical morphospaces (Fig. 7.8). Theoretical morphospaces are produced without any measurement data from real organic form. They are not only independent of existent morphology, they also can be used to produce nonexistent morphology. Such morphospaces exist in the absence of any measurement data (as opposed to empirical morphospaces, which are data dependent; see the discussion in McGhee, 1999, pp. 22–26). As such, theoretical morphospaces are topological spaces (for a discussion of pre-topological spaces in the analysis of biological form, see Stadler *et al.*, 2001; Stadler and Stadler, 2004).

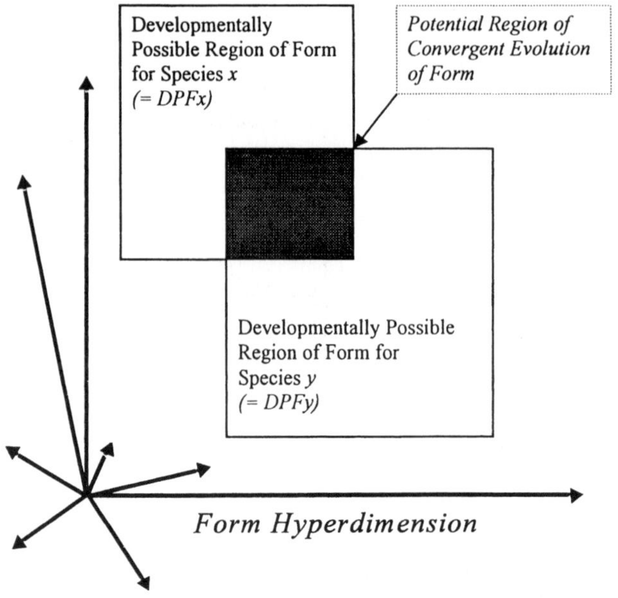

Figure 7.7. Venn diagram of the phenomenon of evolutionary convergence. Convergent form (shaded rectangle) in species x and species y is the set intersection $DPFx \cap DPFy$; that is, these forms are developmentally possible for both species.

The dimensionality of a theoretical morphospace will always be less than the total hyperspace of form dimensions (Fig. 7.8); indeed, one of the goals of theoretical morphospace analysis to to create a morphospace with the lowest dimensionality possible while still capturing the essential aspects of the biological form under consideration (McGhee, 1999, pp. 18–22). However, the boundaries of the theoretical morphospace can hypothetically cross all of the evolutionary constraint boundaries (Fig. 7.8); that is, potentially we can actually map these boundaries within the theoretical morphospace for a given set of actual organisms.

Mapping geometric constraint boundaries

In this section we shall consider two examples where geometric constraint boundaries (Fig. 7.2) have actually been mapped within theoretical morphospaces for two separate groups of organisms, the brachiopods (multicellular animals) and the foraminifera (unicellular eukaryotes). We have encountered the brachiopod shellfish previously in Chapter 5

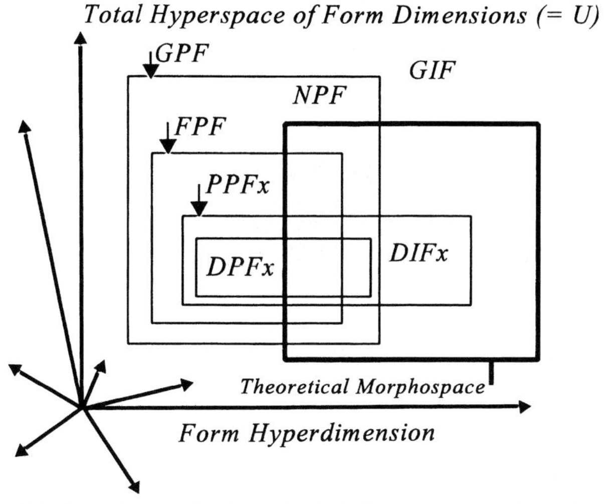

Figure 7.8. Boundaries of a hypothetical theoretical morphospace within the total hyperspace of form dimensions. Note that although the area of the theoretical morphospace (i.e. the dimensionality of the morphospace) is less than that of the total hyperspace of form, the boundaries of the theoretical morphospace may cross all the boundaries of form constraint within the total hyperspace.
Source: From McGhee (2006).

(Figs. 5.3 and 5.4). There we examined a theoretical morphospace designed to simulate the spectrum of possible form that potentially could be found in actual biconvex brachiopods; that is, brachiopods with shells in which both the dorsal and ventral valves are convex (Fig. 5.3).

In their long evolutionary history the brachiopods actually have evolved shell geometries that are much more diverse than just biconvex forms. In the past, particularly in the Palaeozoic, numerous brachiopod species evolved shells in which one of the valves, dorsal or ventral, is essentially flat or lid-like. Other species evolved shells in which one of the valves is concave, where the curvature of the valve is actually recessed within that of the opposing valve. In order to simulate these more complicated shell geometries with the logarithmic spiral model we must abandon the geometric parameter W, the whorl expansion rate, that we used in Chapter 5 to simulate both ammonoid and brachiopod shell forms (for a mathematical demonstration of the truth of this statement, see the discussion in McGhee, 1999, pp. 119–123). Instead, we simply switch to the geometric parameter α, the tangent angle of the logarithmic spiral, that we considered previously in Chapter 4 (Fig. 4.4). A new

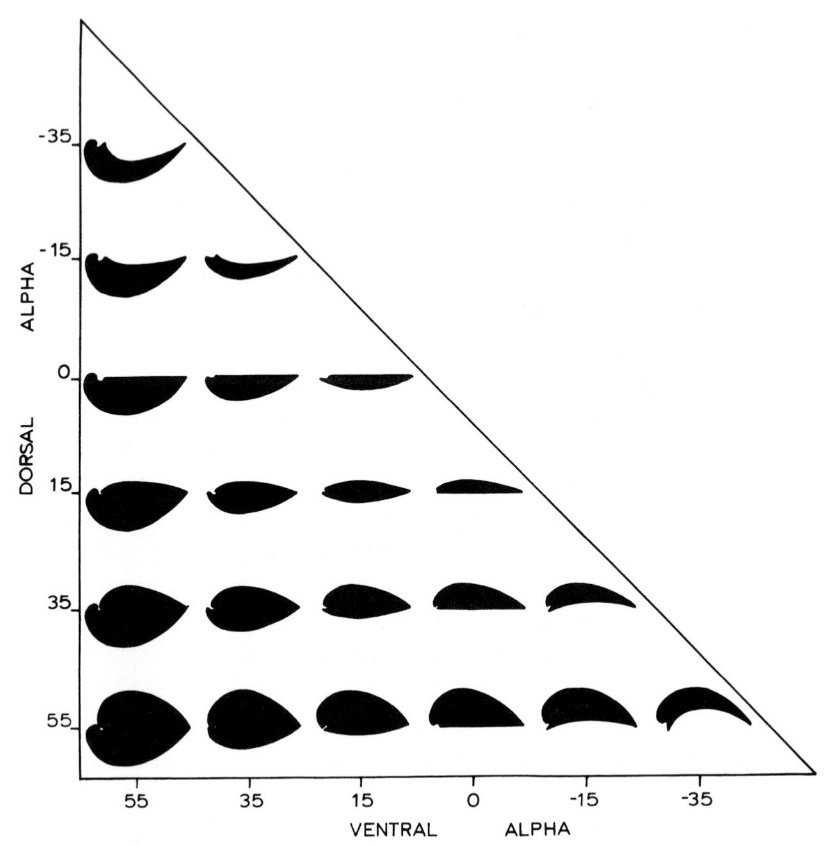

Figure 7.9. A theoretical morphospace of brachiopod shell form, with a computer-produced spectrum of both biconvex and nonbiconvex shell forms. The two morphological-trait dimensions are the tangent angles, α, of the dorsal and ventral valves of the shell.
Source: From McGhee (1999).

theoretical morphospace for brachiopod shell form is illustrated in Figure 7.9, in which not only biconvex shells may be simulated (lower left corner of Fig. 7.9) but also shell geometries with planar to concave dorsal valves (upper left corner of Fig. 7.9), and shell geometries with planar to concave ventral valves (lower right corner of Fig. 7.9).

We have in fact already encountered a geometric constraint boundary in our earlier consideration of brachiopod shell form – that of the boundary between the region of morphospace in which whorl overlap is absent in the valves of the shell, and that of the region of morphospace in which whorl overlap is present (Fig. 5.3). That is, we saw in Chapter 5 that is geometrically impossible to create hinged bivalved shells in which whorl overlap is present in either one or both of the valves.

In the new morphospace of α-parameters (Fig. 7.9), whorl overlap occurs when the α-value of the valve is less than or equal to 70°, if the D-value of the valve is equal to 0.01 (which is typical for brachiopod shells; see Chapter 5). The geometric constraint boundary between the region of possible brachiopod form (geometries with no whorl overlap) and the impossible region of brachiopod form (geometries with whorl overlap) is illustrated in Figure 7.10. The computer simulation in the left margin of Figure 7.10 illustrates the geometry produced when α is equal to 70° in the dorsal valve, and α is equal to 80° in the ventral valve, in a hypothetical brachiopod shell in the whorl overlap region of the morphospace. The computer simulation in the left margin of Figure 7.10 illustrates why it is impossible to articulate two valves together to form a bivalved shell if whorl overlap is present in either valve – the whorls would have to interpenetrate one another; that is, portions of the two separate valves would have to occupy the same space at the same time, which is geometrically impossible.

Interestingly, when we consider the total spectrum of hypothetical brachiopod shell form (Fig. 7.9) another geometric constraint boundary appears that was not apparent when we considered only the potential spectrum of biconvex shell form (Fig. 5.3). That is, *it is impossible to form a bivalved shell if α is negative in both valves.* Both such valves would have convex inner surfaces and concave outer surfaces. The computer simulation in the right margin of Figure 7.10 illustrates the geometry of a hypothetical shell with a dorsal valve having an α-value of −70° and a ventral valve having an α-value of −70°. In essence, the entire region of negative α-coordinates in both valves within the morphospace maps out a region where the hypothetical brachiopod shells that have been turned 'inside-out', and thus are geometrically impossible (Fig. 7.10).

The triangular line surrounding the shaded region labelled 'potential brachiopod morphospace' in Figure 7.10 is an actual geometric constraint boundary, a real demonstration of a boundary between the set of forms *GIF* and *GPF* in Figure 7.2. Thus the techniques of theoretical morphospace analysis can actually map geometric constraint boundaries, rather than simply thinking of such boundaries in a hypothetical and heuristic sense.

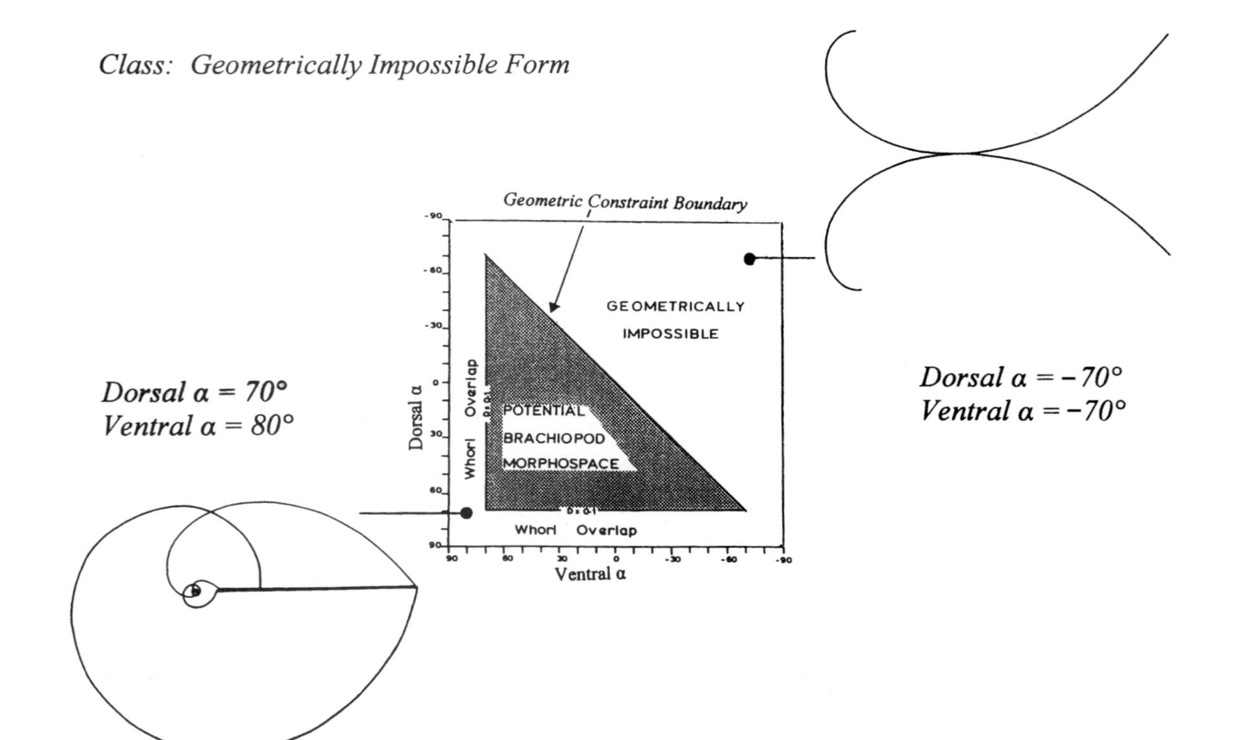

Figure 7.10. Illustration of actual geometric constraint boundaries containing the theoretical morphospace of potential brachiopod form (shaded region) given in Fig. 7.8. Note that the morphospace coordinates dorsal $\alpha = 70°$ vental $\alpha = 80°$, and dorsal $\alpha = -70°$ vental $\alpha = -70°$, produce geometrically impossible shell forms.
Source: Modified from McGhee (1999).

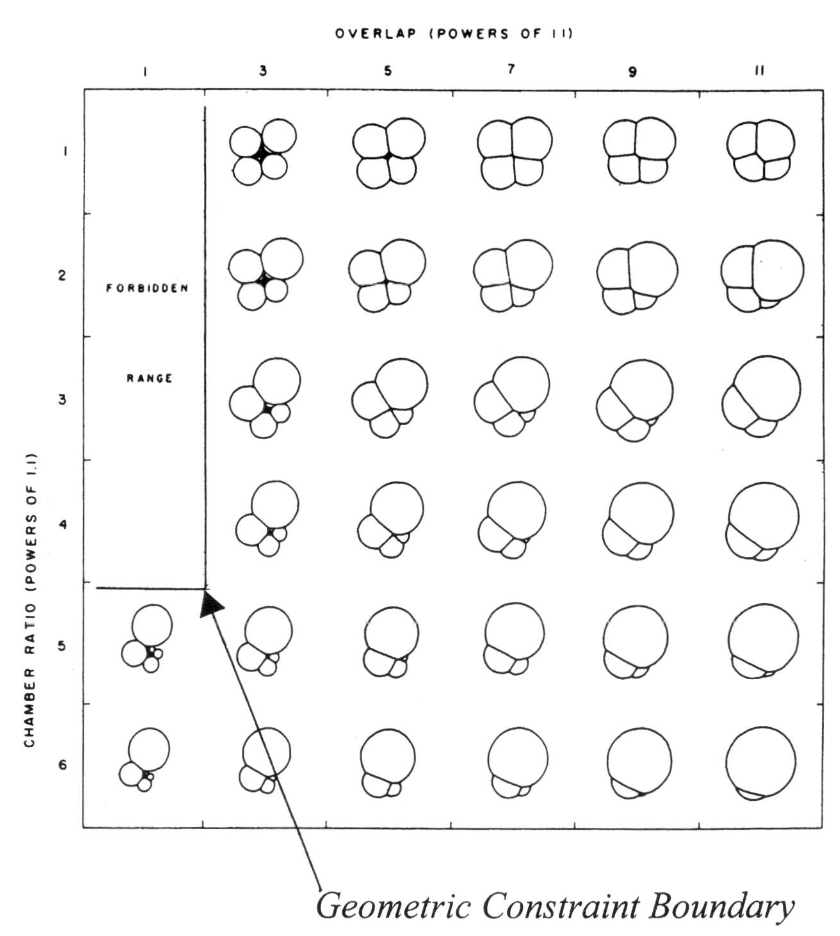

Figure 7.11. Illustration of an actual geometric constraint boundary within a theoretical morphospace of computer-produced foraminiferal test form. The 'forbidden range' region of the morphospace is the region where the dimensional parameter coordinates would produce foraminiferal tests whose chambers no longer overlap, or even touch, one another and are thus geometrically impossible test forms.
Source: Modified from Berger (1969). Copyright © 1969 by SEPM (Society for Sedimentary Geology) and reprinted with permission.

A second quick example of mapping an actual geometric constraint boundary within theoretical morphospace is illustrated in Figure 7.11, from the work of Berger (1969). Figure 7.11 illustrates a theoretical morphospace designed to simulate the spectrum of form found in the tiny shells, called *tests*, of marine foraminifera. The foraminifera are unicellular organisms, passively floating in the world's oceans, that develop intricate test forms composed of small spherical chambers, which grow discretely and are sequentially added to one another. The two morphological-trait parameters of the morphospace are the chamber ratio, which measures the relative placement of the chambers formed in each whorl, and the overlap ratio, which measures the extent to which one chamber overlaps the geometry of the previously formed chamber.

Note that the upper-left region of the morphospace is designated as a 'forbidden range' within the spectrum of hypothetical foraminiferal test forms (Fig. 7.11). Examining the spectrum of computer-simulated foraminiferal tests within the morphospace, you can see that the degree to which the chambers within the tests overlap one another decreases as you move from the lower-right corner of the morphospace to the upper-left corner; that is, as the values of the geometric parameters chamber ratio and overlap ratio decrease. The 'forbidden range' region of the morphospace is the region where the dimensional parameter coordinates would produce foraminiferal tests whose chambers no longer overlap, or even touch, one another. That is, a hypothetical foraminiferal test having a chamber ratio of two, and an overlap ratio of one, would be composed of a series of small chambers that are floating independently in space, unconnected to one another, and thus is a geometrically impossible test form.

These are but two examples of the actual mapping of geometric constraint boundaries in theoretical morphospace using real organisms. Further computer simulation of the evolutionary processes suggest that 'evolutionary mode, tempo and direction may often be the result of constraints inherent in the geometry of the organism' (Swan, 1990, p. 223), yet at present the role of geometric constraint in shaping evolution is largely unrecognized, and unexplored, by evolutionary biologists.

Mapping functional constraint boundaries

Figure 7.3 is a Venn diagram illustrating the relationship between the set of forms that function in nature, *FPF*, and the set of forms that do not, *NPF*. We can take a more spatial representation of these sets by realizing that the forms within these sets not only are characterized by the binary attributes of function versus nonfuction, but that they also have dimensional coordinates (which are unspecified in a simple

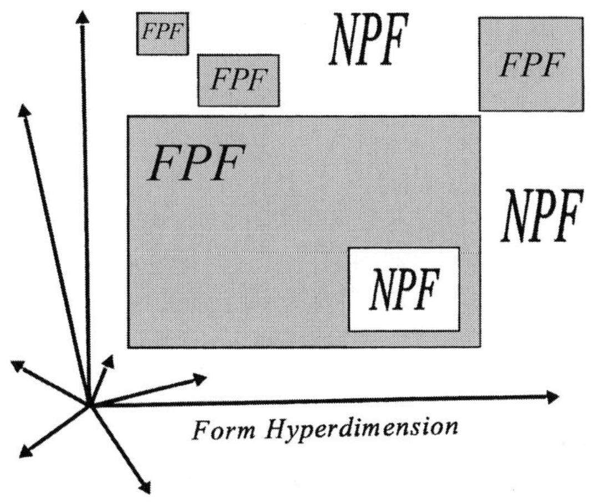

Figure 7.12. A spatial representation of the sets of functional possible form (*FPF*) and nonfunctional possible form (*NPF*) in the geometrically possible region of form (*GPF*) of the total hyperspace of form. Note that the spatial distribution of the two sets of form do not overlap, thus preserving the property $FPF \cap NPF = \oslash$.

Venn diagram representation of the concept of *FPF* versus *NPF*). That is, the forms contained within the sets *FPF* and *NPF* do not necessarily occur in a single region of the total hyperspace of form, and that those regions are not necessarily contiguous. Figure 7.12 gives a hypothetical spatial realization of the Venn diagram in Figure 7.3. Note the nonoverlapping regions of functional possible form and nonfunctional possible form in Figure 7.12; that is, the spatial representation preserves the set configurations:

$$FPF \cup NPF = \{f \mid f \in FPF \quad \text{or} \quad f \in NPF\} = GPF,$$
$$FPF \cap NPF = \{f \mid f \in FPF \quad \text{and} \quad f \in NPF\} = \oslash,$$

that we outlined earlier in the chapter. Viewing Figure 7.12 from a landscape or geographic perspective, we have a large continent of functional possible form, and a few smaller islands of functional possible form, sitting in an ocean of nonfunctional possible form in the hyperspace region of geometrically possible form. Within the large region of functional possible form exists a large lake of nonfunctional possible form, totally surrounded by the region of functional possible form. Of course, these regions could themselves be expressed precisely as subsets of either *FPF* or *NPF* by adding the additional information of

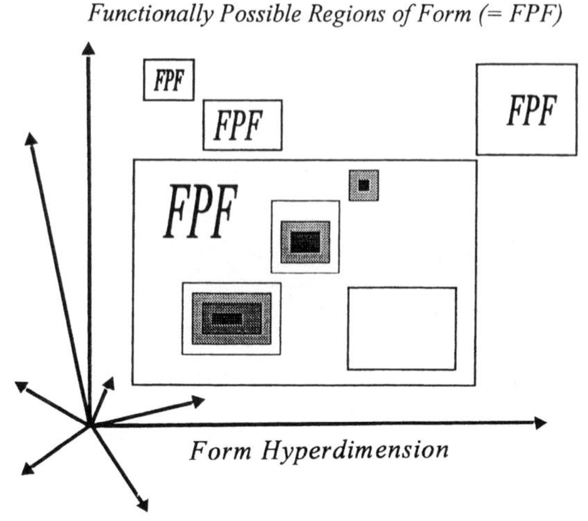

Figure 7.13. A spatial representation of differing degrees of the adaptive value of forms within the region of functional possible form (*FPF*) Dark-shaded rectangles represent the *FPF* regions with the best adapted forms, lighter-shaded regions contain less well-adapted forms, and nonshaded regions contain the least adapted forms within the region of functional possible forms.

their spatial coordinates to their subset designations, but that is not the point of the present discussion.

Although it is theoretically possible that all of the forms found in the spatial regions of functional possible form are of equal adaptive value, we know that that possibility is highly unlikely. Some of the functional possible forms will function better, having a higher adaptive value, than others. Continuing our landscape reasoning, we can contour the adaptive value of differential forms within the spatial regions of functional possible form. Thus, in Figure 7.13, we see that three mountains exist on the continent of functional possible form, surrounded by lowlands and low-lying islands of functional possible form. In Figure 7.13 we have simply reinvented the adaptive landscape, where the adaptive peaks and slopes give the set of forms *FPF* and the flat adaptive plane contains the set of forms *NPF* (Fig. 7.14).

In Chapters 5 and 6 we considered multiple examples of the actual spatial mapping of adaptive surfaces and of the actual spatial boundaries of the frequency of form found in real organisms, in theoretical morphospaces. We examined real adaptive peaks occupied by real ammonoids, brachiopods, terrestrial plants, bryozoans and so on.

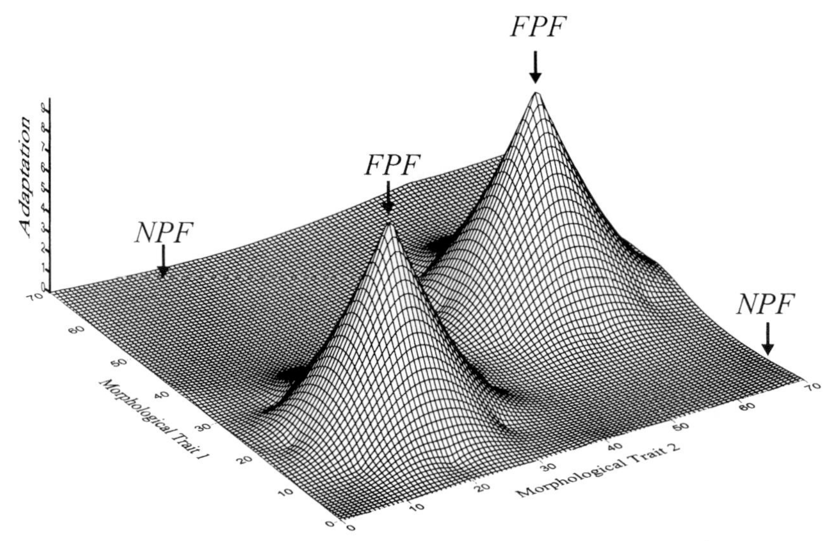

Figure 7.14. An adaptive landscape where the adaptive peaks and slopes give the set of forms *FPF* and the flat adaptive plane contains the set of forms *NPF*.

Very importantly, however, we saw that the techniques of theoretical morphospace analysis actually allow us also to create and examine *nonexistent forms*, and that the geometries and functional properties of these nonexistent forms can be used to analyse nonfunctional spatial regions of empty morphospace. Thus the techniques of theoretical morphospace analysis can actually map functional constraint boundaries, a real demonstration of the boundaries between the set of forms *NPF* and *FPF* in Figure 7.12, as well as the mountains and lowlands seen within *FPF* in Figures 7.13 and 7.14.

Analysing intrinsic constraints

The analytic techniques of theoretical morphology are particularly powerful when applied to extrinsic constraints, those of geometry and function. The analysis of intrinsic constraints, those imposed by the biology of specific organisms, is more difficult and often is more the assessment of a series of probabilities within theoretical morphospace, rather than a deterministic demonstration of geometrically possible versus impossible, or functional versus nonfuctional, regions of morphospace.

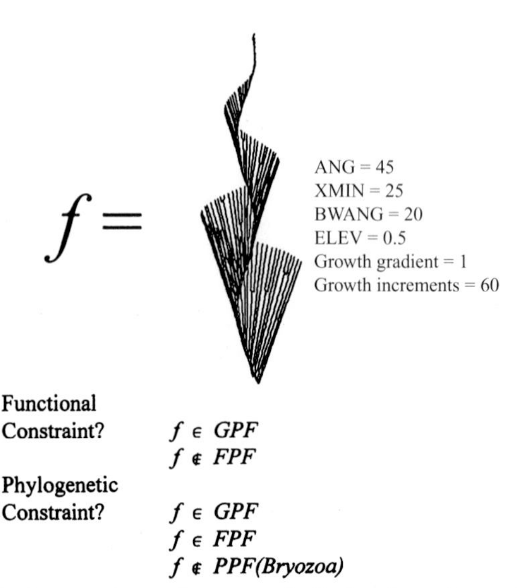

Figure 7.15. A nonexistent bryozoan helical colony form, f, from the theoretical morphospace of helical colony form discussed in Chapter 6. The computer simulation illustrates the fact that the form is geometrically possible (i.e., $f \in GPF$). The nonexistence of the form in nature may be because it is maladaptive (functional constraint, $f \notin FPF$) or because the Bryozoa do not possess the genetic coding necessary to develop the form (phylogenetic constraint, $f \notin PPF(Bryozoa)$).

We shall consider two actual examples of the analysis of the question of intrinsic constraint in this section. First, let us return to the helical bryozoan colonies that we considered in Chapter 6. We saw that is it possible to create a theoretical morphospace of helical colony form (Fig. 6.2) that allowed us to computer simulate actual geometries found in existent bryozoan colonies (Figs. 6.1 and 6.3), as well to simulate helical colony geometries that have never been evolved by bryozoans (Fig. 6.4), even though they are geometrically possible.

Figure 7.15 illustrates one of the forms, f, that clearly belongs to the set of geometrically possible forms, GPF, yet nevertheless has never been evolved in nature and thus is located in a region of morphospace that is empty of bryozoans (Fig. 6.4). Why have the bryozoans never evolved this elegant form? Two possibilities immediately arise; first, the form may be nonfunctional and its nonexistence is due to extrinsic functional constraint. That is, the form is a member of the set of forms GPF that do not belong to the set FPF:

$$GPF \backslash FPF = \{f \mid f \in GPF, f \notin FPF\}.$$

Alternatively, we can hypothesize that the form functions just fine, but that it is impossible for the bryozoans to evolve it due to constraints intrinsic to the bryozoans: either phylogenetic constraint, in that the genetic coding necessary for producing the form is not present in the bryozoan genome:

$$FPF \backslash PPF(Bryozoa) = \{f | f \in FPF, f \notin PPF(Bryozoa)\},$$

or that the genetic coding is present, but that it is developmentally impossible for the bryozoans to develop the form:

$$PPF(Bryozoa) \backslash DPF(Bryozoa) = \{f | f \in PPF(Bryozoa), f \notin DPF(Bryozoa)\}.$$

The intrinsic constraint hypothesis for the nonexistence of the form f in Figure 7.15 can be refuted in two ways, one deterministic and the other probabilistic. Functional analyses demonstrate that the nonexistent form f possesses aspects of geometry that are unrealistic as fluid-filtering geometries for filter-feeding organisms (as discussed in Chapter 6; see McGhee and McKinney, 2000; McKinney and McGhee, 2003). This is the easiest approach to refuting a hypothesis of intrinsic constraint: by demonstrating that the constraint is extrinsic.

However, for the sake of heuristic argument, let us imagine at this point that we cannot demonstrate in this case that the form f does not belong to the set of *FPF*. We can still argue that the nonexistence of form f is not due to phylogenetic constraint even in the absence of the deterministic proof that the form is nonfunctional.

Phylogenetic constraints are best analysed within the framework of phylogenetic hypothesis construction (McKitrick, 1993; Schwenk, 1995). Cladistic (phylogenetic) analyses reveal that helical colony geometries have convergently, independently, evolved (see Chapter 6) in widely separate regions of the total cladogram for the Phylum Bryozoa (McKinney and McGhee, 2003). That is, the extinct species of the helical bryozoan genera *Archimedes* and *Chrisidmonea* belong to two separate Orders within the Class Stenolaemata, thus they are only distantly related to one another. The extant species of the helical bryozoan genera *Bugula* and *Retiflustra* are even more distantly related, in that they belong to an entirely different Class of bryozoans, the Gymnolaemata. Thus the genetic coding permitting the development of helical colony forms must be present near the very base of the cladogram of the bryozoans, near the evolution of the phylum Bryozoa itself in the early Palaeozoic, and thus had to be present some 495 million years ago.

Thomas and Reif (1993) have argued that, given enough evolutionary time and an extremely large number of evolutionary experiments, the evolutionary discovery of the spectrum of functional morphologies obtainable by a group of organisms is essentially inevitable. The repeated convergent evolution of certain helical morphologies by the bryozoans (Fig. 6.3), but not others (Fig. 6.4), corroborates this argument. On probabilistic grounds one can further argue that if the observed nonexistent colony form *f* (Fig. 7.15) was functional, the bryozoans would surely have discovered it since the genetic coding necessary for producing helical colony forms has been present for some 495 million years of bryozoan evolution. That is, its nonexistence is not due to intrinsic phylogenetic or developmental constraints within the phylum Bryozoa.

Are we to conclude then that phylogenetic or developmental constraints do not in fact exist? Not at all – let us consider a second example of the analysis of empty morphospace with respect to the question of intrinsic constraint. In Chapter 6 we considered the morphological effects of biodiversity crises in the evolution of the ammonoids in theoretical morphospace. The ammonoids survived the Late Devonian biodiversity crisis (Fig. 6.8) and the End-Permian biodiversity crisis (Fig. 6.9), but they did not survive the End-Cretaceous biodiversity crisis, becoming totally extinct some 65 million years ago. However, a distantly related group of swimming cephalopods with external shells, the nautilids, did survive the End-Cretaceous event and are alive today.

An early study that examined the differential evolutionary fates of the ammonoids and nautilids within theoretical morphospace is that of Ward (1980). In the Cretaceous, the ammonoids and nautilids inhabited distinctly different regions of morphospace, with very little overlap in the spectrum of morphologies found in the two groups of swimming cephalopods (Fig. 7.16). Obviously, as the ammonoids go extinct at the end of the Cretaceous, the region of morphospace formerly inhabited by these animals is vacant in the early Cenozoic. Of interest is the region of morphospace inhabited by the nautilids during the subsequent Cenozoic, following the demise of the ammonoids (Fig. 7.16).

Ward (1980) found that a very interesting shift in the frequency distribution of nautilid morphologies occurs in theoretical morphospace following the extinction of the ammonoids (Fig. 7.16). Values of *D*

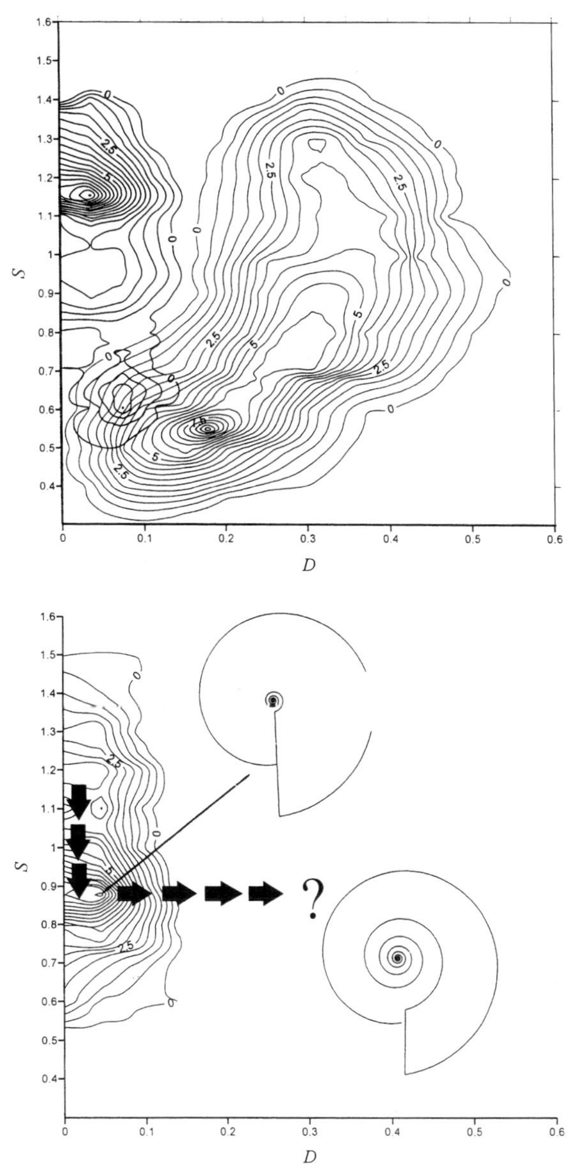

Figure 7.16. Contoured frequency distribution of shell forms found in Cretaceous ammonoids and nautilids (top figure), and in Cenozoic nautilids (bottom figure), in Raup's theoretical morphospace of cephalopod shell forms discussed in Chapters 4 and 5. In the top figure, the nautilid frequency distribution is on the left, and ammonoid shell forms are the much larger frequency distribution on the right. Following the extinction of the ammonoids, the nautilids shifted their distribution downwards from a peak at $S = 1.15$ to $S = 0.9$, shown by the vertical vectors in the bottom figure. The fact that Cenozoic nautilids never invaded the large empty region of the morphospace, previously occupied by ammonoids, is indicated by the horizontal vectors and question mark.
Source: Data from Ward (1980).

remain about the same, but many more nautilids evolved with lower values of *S* in the Cenozoic than were present in the Cretaceous. The peak, or mode, of the morphological frequency distribution also exhibits this shift, moving from a position of *S* equal to about 1.15 in the Cretaceous to a position of *S* around 0.9 in the Cenozoic (Fig. 7.16). What does this mean? More nautilids have moved into the region of morphospace that had previously been occupied by the ammonites! Ward (1980, p. 32) does not think that this pattern is coincidence, and suggests that 'the terminal Cretaceous extinction of ammonites may have opened up new opportunities for nautilid evolution during the Tertiary, because Tertiary nautilids are dominated by moderately compressed, hydrodynamically efficient shell shapes which were rarely present among Jurassic and Cretaceous nautilids, but common among ammonites.' Ward (1980) noted, however, that the region of ammonite morphospace which has been successfully invaded by nautilids is very small compared to the former total range of form exhibited by the ammonites.

Now let us consider this evolutionary phenomenon from an opposite perspective. In fact, the great majority of theoretical morphospace once occupied by the ammonoids (see Fig. 7.16) was vacated when the ammonoids became extinct, and remains empty today. Why? If the nautilids were able to successfully invade the lower-*S* regions of the morphospace once the ammonoids became extinct, why have they not been able to also invade the vastly larger higher-*D* regions of empty morphospace that remain empty to this day?

On probabilistic grounds one can argue that the inability of the nautilids to evolve morphologies present in the empty region of morphospace is due to intrinsic phylogenetic constraint. First, it is clear that shell forms in this region of the morphospace belong to the set of functional possible forms, *FPF*, for swimming ectocochleate cephalopods as they were successfully utilized by the ammonoids for the greater part of the Palaeozoic and the entire Mesozoic. Second, the nautilids have had 65 million years to evolve these forms, yet they have not. Reasoning from Thomas and Reif (1993), we know the morphologies in the empty region of morphospace are functional and we know the nautilids have had a long span of evolutionary time to discover them, thus we should conclude that the re-evolution of these morphologies would be inevitable. Thus it is extraordinary that this region of the morphospace continues to remain empty, and argues for the action of another type of constraint other than the extrinsic constraint of function.

The extinct ammonoids had long, eel-like bodies, encased in narrow shells with numerous whorls. Often their eel-like bodies extended in a narrow tube a full revolution back into the shell, as we can see in their fossil shells that the living chamber is typically over 360° in length (Saunders, Work and Nikolaeva, 2004). In contrast, the nautilids have short, thick bodies, encased in broad shells with few whorls. The living chamber of the shell containing their bulbous bodies is typically very short, less than 180° in length. The phylogenetic legacy of the bodies and organ systems of these two groups of cephalopods is very different. It is also very ancient: you have to go back in time to the early Palaeozoic, over 400 million years, to find the most recent common ancestor of these two groups of superficially similar swimming cephalopods.

The probability is thus high that the nautilids simply do not possess the genetic coding necessary to produce the shell morphologies found in the empty region of the morphospace (Fig. 7.16). These shell geometries are fundamentally at variance with the shape and organ distribution of the nautilid body itself, and the 400-million-year-old phylogenetic legacy of the nautilids would have to undergo a radical reorganization in order to allow the animals to utilize these shell forms. These shell morphologies are unattainable by the nautilids due to constraints imposed by the specific biology of the nautilids themselves; that is, by intrinsic constraint.

Last, the concept of empty morphospace and its analysis can even be extended to the field of community ecology. Mack (2003) has conducted an interesting comparative analysis of plant communities in which he argues that the phenomenon of 'absent life forms' in some communities is due to phylogenetic constraint, in that the resident plant species are unable to produce these ecological (and morphological) types. He has further argued that the relative success of invasive, alien species in many cases is a function of their ability to produce the forms absent in the native plant communities. Thus the analytic techniques of theoretical morphology might be applied to the analysis of the spectrum of ecomorphologies present in communites, and that the detection of 'absent life forms' within a community may provide a clue for predicting the potential success of alien invasive species in such communities (Mack, 2003).

Modelling the evolution of intrinsic constraints

The boundaries of the extrinsic constraints of geometry and function in form hyperspace are absolute and do not vary with time. That is, the laws of geometry and the physics of flying or swimming are the same today as they were 100 million years ago. This is not to say that local physical changes in the universe do not occur, changes that may alter the local conditions determining functional versus nonfunctional form. For example, the atmosphere of Mars four thousand million years ago was much thicker and denser than at present. Thus the type of morphology that could have functioned as a powered flying organism on Mars in the early history of the planet cannot possibly be functional in the thin atmosphere present on the planet today; that is, much more wing surface area would be necessary to produce lift in the current Martian atmosphere. But within the period of time in the evolution of the universe within which the physical conditions of the universe are such that life can exist, the boundaries of the total sets of form *GIF* versus *GPF*, and *FPF* versus *NPF*, do not change within the universal set of form *U*.

This is not true of the boundaries of intrinsic constraints. Intrinsic constraints are imposed by the specific biology of specific organisms and, as organisms evolve in time, the boundaries of their intrinsic constraints must evolve and change as well within the total sets of geometrically possible form, *GPF*, and functional possible form, *FPF*.

As we did previously in our considerations of possible spatial distributions of the boundaries of the sets *FPF* and *NPF* (Figs. 7.12 and 7.13), let us consider possible spatial and temporal distributions of the boundaries of the phylogenetically possible set of forms for a group of organisms of lineage x, or $PPF(x)$, as those boundaries might change in time, t. If we consider three time intervals, $t1$, $t2$ and $t3$, we then can designate three sets of phylogenetically possible form for lineage x:

$$PPF(x, t1) = \{f \mid f = \text{phylogenetically possible forms for lineage } x \text{ at } time\ 1\},$$
$$PPF(x, t2) = \{f \mid f = \text{phylogenetically possible forms for lineage } x \text{ at } time\ 2\},$$
$$PPF(x, t3) = \{f \mid f = \text{phylogenetically possible forms for lineage } x \text{ at } time\ 3\}.$$

Let us now explore some of the possible evolutionary changes that might occur in the boundaries of these three sets. In this modelling section, we shall consider only the possible changes in the boundaries of *PPFx* that theoretically might occur within the boundaries of the larger set of functional possible form, *FPF*. Although it is theoretically possible that the set *PPFx* could have boundaries that hypothetically cross the boundaries of both functional and nonfunctional possible form (*FPF* and *NPF* in Fig. 7.4), and even the boundaries of geometrically possible and impossible form (*GPF* and *GIF* in Fig. 7.4), it is only within the set intersection $PPFx \cap FPF$ that *existent form* can potentially be found, depending upon the boundaries of the set of developmentally possible forms for lineage *x*, *DPFx* (Fig. 7.6), a condition that we shall consider in more detail in Chapter 8. A key feature of the concept of phylogenetic constraint is there may exist functional possible form that cannot be attained by a group of organisms purely due to constraints imposed by its own phyletic legacy, thus from a theoretical morphospace analytical perspective, we are looking for empty regions of *FPF* morphospace that might be due to the action of this constraint.

One common evolutionary phenomenon is that of diversification, where an original ancestral species radiates into a clade of new species descendants. We can model morphological diversification as the process of the expansion of the boundaries of the phylogenetically possible region of form within form hyperspace for a genetic lineage with time (Fig. 7.17). In Figure 7.17, the region of form, the spectrum of morphologies, that are phylogenetically possible for lineage *x* at time *t1* is quite small. With the evolution of new genetic coding with time, this initial region of phylogenetically possible form is modelled as expanding within the form hyperspace. That is, at time *t3* the lineage *x* possesses the genetic coding needed to produce morphologies that were not possible for that lineage at time *t1*; regions of morphospace that were outside the boundaries of the set of forms *PPF(x, t1)* are now within the boundaries of the set of forms *PPF(x, t3)*. A hypothetical phylogenetic tree for lineage *x* is given in Figure 7.17, which represents a possible observable pattern of speciation in the fossil record that corresponds to the modelled expansion of the lineage's boundaries of phylogenetically possible form within the form hyperspace. Note that in this model of diversification, the genetic legacy of the lineage is conserved, such that the sets of possible form are nested within one another:

$$PPF(x, t1) \subset PPF(x, t2) \subset PPF(x, t3).$$

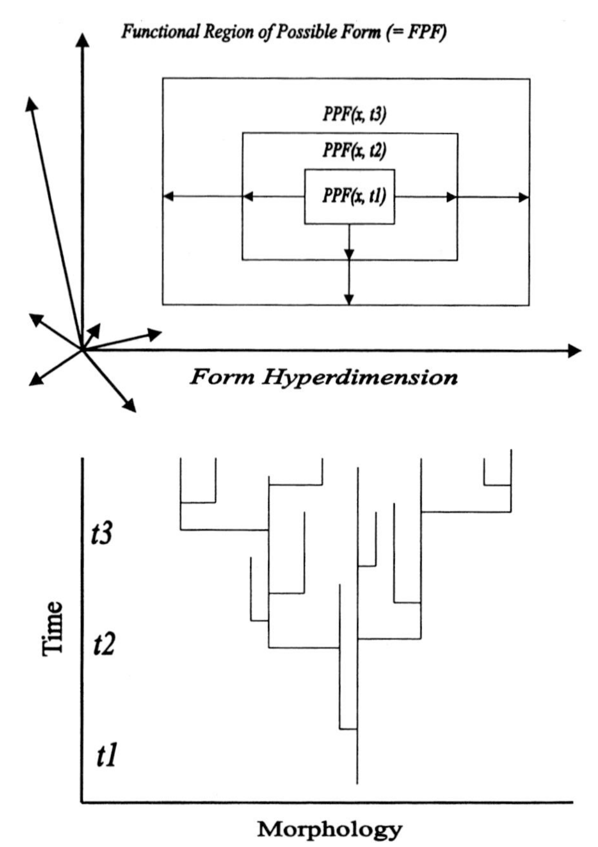

Figure 7.17. Evolutionary expansion. Upper figure: spatial representation of the hypothetical evolutionary expansion of the region of phylogenetically possible form for lineage x in the region of functional possible form (FPF), from time-interval one ($PPF(x, t1)$) to time-interval three ($PPF(x, t3)$). Lower figure: a hypothetical phylogenetic tree of speciation events within lineage x, where the morphological diversification of the lineage's species corresponds to the expansion of the lineage's $PPFx$ in the form hyperspace (upper figure).

An opposite phenomenon to that of diversification is a genetic bottleneck, in which genetic legacy is lost and the spectrum of morphologies seen within a lineage becomes more restricted with time. We can model a genetic bottleneck as a contraction of the region of phylogenetically possible form for lineage x within form hyperspace with time (Fig. 7.18). Regions of morphospace that were within the boundaries of the set of forms $PPF(x, t1)$ are now outside the boundaries of the set of forms $PPF(x, t3)$, and morphologies that were once present within

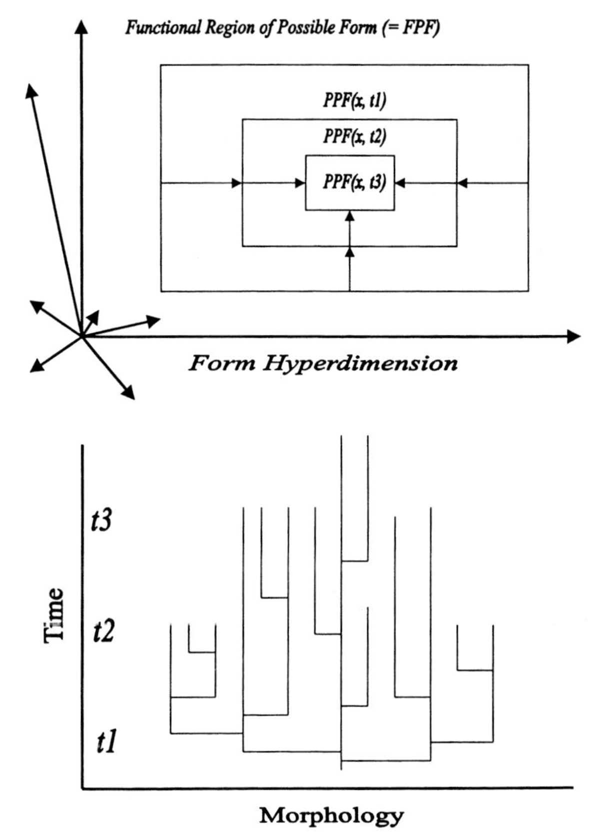

Figure 7.18. Evolutionary contraction. Upper figure: spatial representation of the hypothetical evolutionary contraction and bottleneck of the region of phylogenetically possible form for lineage x in the region of functional possible form (*FPF*), from time-interval one (*PPF(x, t1)*) to time-interval three (*PPF(x, t3)*). Lower figure: a hypothetical phylogenetic tree of speciation events within lineage x, where the loss in morphological diversity of the lineage's species corresponds to the contraction of the lineage's *PPFx* in the form hyperspace (upper figure).

the lineage are no longer possible for that lineage to produce as the genetic coding necessary to produce them has been lost. A hypothetical phylogenetic tree for lineage x, corresponding to the modelled contraction of the lineage's boundaries of phylogenetically possible region of form within the form hyperspace, is given in Figure 7.18. Note that in this model of a genetic bottleneck, genetic legacy and morphological diversity is lost, such that:

$$PPF(x, t1) \supset PPF(x, t2) \supset PPF(x, t3).$$

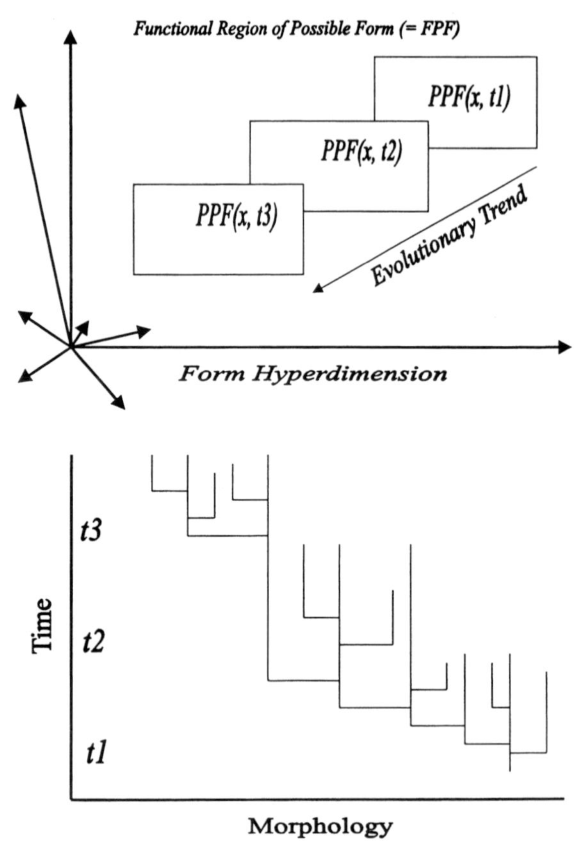

Figure 7.19. Evolutionary trends (Part 1). Upper figure: spatial representation of a hypothetical evolutionary trend in the location of the region of phylogenetically possible form for lineage x in the region of functional possible form (*FPF*), from time-interval one (*PPF(x, t1)*) to time-interval three (*PPF(x, t3)*). Lower figure: a hypothetical phylogenetic tree of speciation events within lineage x, where the trend in the morphologies of the lineage's species corresponds to the trend in the location of the lineage's *PPFx* in the form hyperspace (upper figure).

In both the model of morphological diversification (Fig. 7.17) and diversity loss (Fig. 7.18), the temporal boundaries of the phylogenetically possible region of form for lineage x are shown as centred and symmetrical. We can model the phenomenon of an evolutionary trend by introducing asymmetry into the models. In Figure 7.19, the boundaries of the phylogenetically possible region of form for lineage x are shown as moving from one region in the form hyperspace to another, and not just symmetrically expanding or contracting. A hypothetical phylogenetic tree for lineage x, representing a possible observable pattern of speciation in the fossil record that corresponds to the modelled movement of the lineage's boundaries of phylogenetically possible form within the form hyperspace, is given in Figure 7.19.

The model of an evolutionary trend shown in Figure 7.19 entails both the evolution of new genetic novelties and the loss of genetic legacy. Regions of morphospace that were outside the boundary of the set of forms $PPF(x, t1)$ are now within the boundary of the set of forms $PPF(x, t3)$; that is, the lineage at time $t3$ now possesses the genetic coding to produce morphologies that were impossible for the lineage to produce at time $t1$. On the other hand, note that the model also indicates that regions of morphospace that were within the boundary of the set of forms $PPF(x, t1)$ are now outside the boundary of the set of forms $PPF(x, t3)$, and morphologies that were once present within the lineage are no longer possible for that lineage to produce as the genetic coding necessary to produce them has been lost (as seen in the hypothetical phylogenetic tree for the lineage). That is, the evolutionary trend is modelled as the sequential overlapping of the temporal sets of form such that:

$$PPF(x, t1) \cap PPF(x, t2) \neq \emptyset \text{ and } PPF(x, t2) \cap PPF(x, t3) \neq \emptyset,$$

but that eventually the trend produces the condition:

$$PPF(x, t1) \cap PPF(x, t3) = \emptyset,$$

where the original set of form $PPF(x, t1)$ is no longer available and genetic legacy has been lost. Only set $PPF(x, t3)$ exists at time $t3$.

The loss of genetic legacy is not a necessary condition of modelling an evolutionary trend. In Figure 7.20 an evolutionary trend is modelled in which genetic legacy is conserved, in that the trend is produced by the asymmetrical expansion of the boundaries of phylogenetically possible form for lineage x in form hyperspace, rather than the symmetrical expansion of these boundaries as modelled in Figure 7.17. As in the model of morphological diversification, the sets of form shown in Figure 7.20 meet the condition:

$$PPF(x, t1) \subset PPF(x, t2) \subset PPF(x, t3).$$

A hypothetical phylogenetic tree for lineage x, representing a possible observable pattern of speciation in the fossil record that corresponds to the modelled asymmetrical expansion of the lineage's boundaries

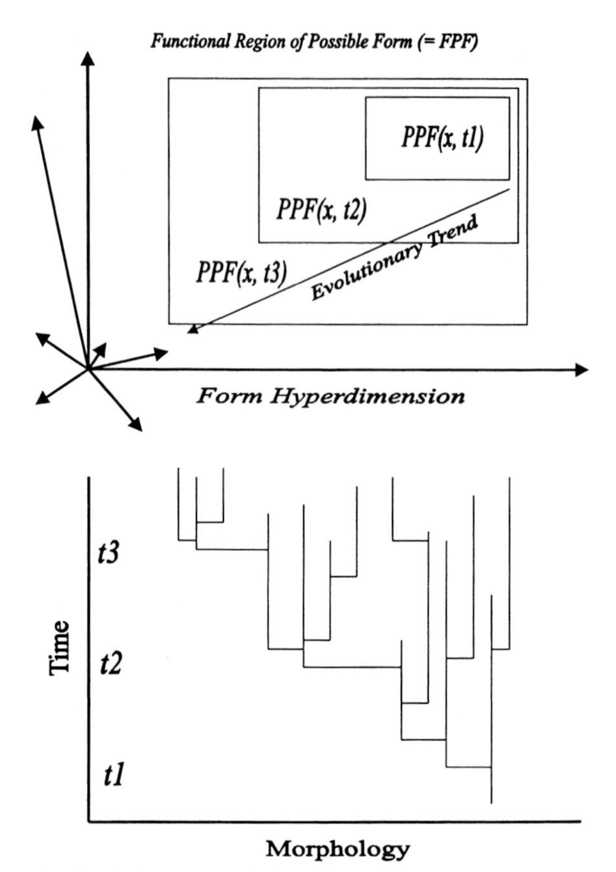

Figure 7.20. Evolutionary trends (Part 2). Upper figure: spatial representation of a hypothetical evolutionary trend produced by the asymmetrical expansion of the region of phylogenetically possible form for lineage x in the region of functional possible form (*FPF*), from time-interval one (*PPF(x, t1)*) to time-interval three (*PPF(x, t3)*). Lower figure: a hypothetical phylogenetic tree of speciation events within lineage x, where the trend in the morphologies of the lineage's species corresponds to the trend in the asymmetrical expansion of the lineage's *PPFx* in the form hyperspace (upper figure).

of phylogenetically possible form within the form hyperspace, is given in Figure 7.20. To a palaeontologist, the phylogenetic tree illustrated in Figure 7.19 looks much more realistic than that shown in Figure 7.20. In Chapter 3 we considered the morphological evolution of the horse family, which the palaeontologist George Gaylord Simpson modelled in an adaptive landscape format back in 1953 (Fig. 3.5). Modern members of the horse lineage (Equinae) possess only a single toe in the forelimb and are quite large animals, whereas ancestral animals in the horse lineage (Hyracotheriinae) back in the Eocene had four toes in the forelimb and were very small. There exist no small, four-toed members of the horse family today, thus does this mean:

$$PPF(Equidae,\ Today) \cap PPF(Equidae,\ Eocene) = \emptyset,$$

and that the genetic coding necessary to produce small, four-toed animals has been lost, as modelled in Figure 7.19? Certainly the observed phylogenetic tree of morphological evolution within the horse family is much more similar to that shown in Figure 7.19 than 7.20.

Or has the genetic legacy of the ancient Eocene members of the horse family been conserved in modern day horses? Charles Darwin himself was fascinated with morphological atavisms, which he termed 'reversions to long-lost characters'. He was particularly interested in the mysterious appearance of stripes of different fur-colour on the legs and shoulders of many breeds of modern living horses, ponies and asses and speculated in his book, *On the Origin of Species*, that the original ancestor for these equids was striped like the zebra and that the random reappearance of stripes in modern equids reflected a 'reversion' to this primitive ancestral morphology: 'For myself, I venture confidently to look back thousands on thousands of generations, and I see an animal striped like a zebra, but perhaps otherwise very differently constructed, the common parent of our domestic horse, whether or not it be descended from one or more wild stocks, of the ass, the hemionus, quagga, and zebra' (Darwin, 1859, p. 167). In essence, before the science of genetics was founded, Darwin (1859) considered these morphological reversions to be evidence for the conservation of genetic legacy within the horse family.

In this alternative view, the absence of small, four-toed members of the equid lineage today is due to functional constraint, rather than phylogenetic constraint. Under this hypothesis, the ancient Eocene equid browsers have been ecologically displaced by other browsing animals during the span of the Cenozoic, transforming the phylogenetic tree shown in Figure 7.20 into the phylogenetic tree shown in Figure 7.19 by functional constraint. That is, the modern day equids still possess the genetic legacy to produce Eocene-type morphologies, but these ancient browsing-equid morphologies are no longer functional in the modern world.

Could modern horses re-evolve four-toed miniature morphologies? This is a question that ultimately can only be answered by the detailed analysis of the genome of modern horses; it is a theoretical question for the empiricist and experimentalist to consider. From a theoretical perspective, is the re-evolution of four-toed miniature equid morphologies any less probable than the re-evolution of fish-like fins for swimming in land-dwelling tetrapods? Yet we know that the latter has indeed occurred, twice no less, in the convergent re-evolution of such morphologies in ichthyosaurs and porpoises (Fig. 3.1).

Let us consider one last theoretical morphospace model of a hotly-contested evolutionary phenomenon proposed by the palaeontologist Stephen Jay Gould: the 'pruned bush' model for the evolution of morphological disparity. In this model, Gould (1989, 1991) proposed that the spectrum of morphological disparity, the numbers of different types of morphologies present, in Cambrian arthropods was much greater than that seen in the Arthropoda today. With the passage of time, he proposed that more and more ancestral morphologies have been lost, and thus that the phylogenetic tree for the Arthropoda resembles a bush, rather than tree. That is, rather than a tree with ever more branches, and hence morphological diversity, increasing and expanding from the distant past to the present, the tree of life for the Arthropoda is more like a bush that has been progressively pruned, with many branches at its base back in the Cambrian but with only a few extending upward in time to the present.

In order to produce such a model in a theoretical morphospace format, one can consider the case where the phylogenetically possible region of form for a lineage actually fragments in form hyperspace with time, rather than expanding (Fig. 7.17), contracting (Fig. 7.18), or shifting (Figs. 7.19 and 7.20). In such a model, genetic legacy is of necessity lost (Fig. 7.21), such that:

$$PPF(x, t1) \supset PPF(x, t2) \supset PPF(x, t3),$$

as in Figure 7.18. However, in a spatial represention, the sets of phylogenetically possible form *PPF(x, t2)* and *PPF(x, t3)* are composed of different spatial subsets of form within the form hyperspace that are not spatially contiguous (Fig. 7.21), similar to our previous spatial consideration of the single sets of functional and nonfunctional possible form, *FPF* and *NPF*, that may have multiple spatial subsets in form hyperspace (Fig. 7.12).

Figure 7.21 illustrates the resultant 'pruned bush' hypothetical phylogenetic tree of speciation events within lineage *x*, where the loss of ancestral morphologies, and the fragmentation of the morphological

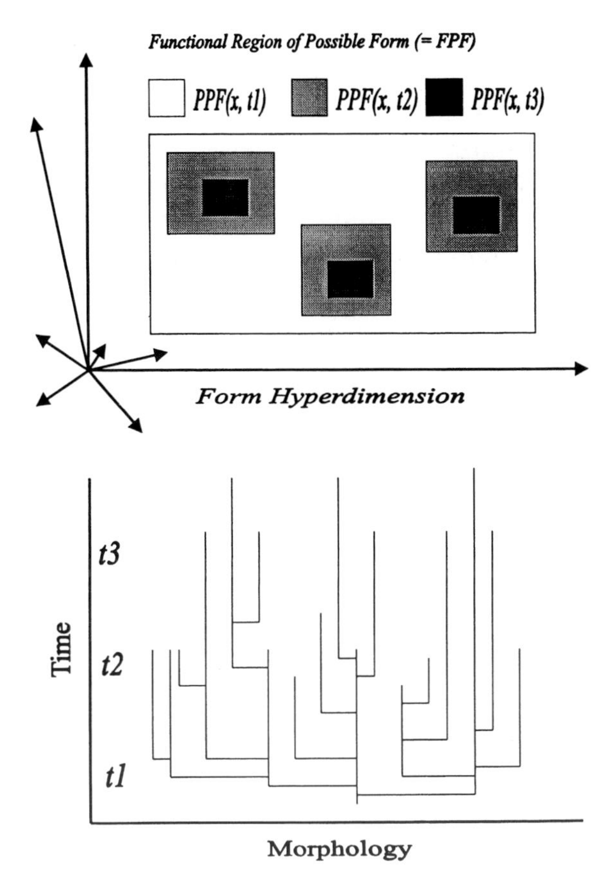

Figure 7.21. Evolutionary fragmentation. Upper figure: spatial representation of the hypothetical fragmentation of the region of phylogenetically possible form for lineage *x* in the region of functional possible form (*FPF*), from time-interval one (*PPF(x, t1)*) to time-interval three (*PPF(x, t3)*). Lower figure: a hypothetical phylogenetic tree of speciation events within lineage *x*, where the loss of morphologies, and the fragmentation of the morphological distribution of the lineage's species, corresponds to the fragmentation of the lineage's *PPFx* in the form hyperspace (upper figure).

distribution of the lineage's species, corresponds to the fragmentation of the lineage's *PPFx* in the form hyperspace. In essence, rather than experiencing a single genetic bottleneck (Fig. 7.18), the evolving morphological lineage *x* is broken up by a series of bottlenecks (Fig. 7.21). The applicability of such a model to the evolution of life, or its frequency of occurrence in the evolution of life, remains controversial to this day (see discussions in Conway Morris, 1998 and McGhee, 1999).

Evolutionary constraint and the phenomenon of convergent evolution

The analysis of the astonishing number of convergently evolved morphologies seen in animals and plants is a major research focus in the evolutionary sciences, but the further implications of the phenomenon is also the subject of philosophical, even religious, discussion (Conway Morris, 2003). For example, the best-known evolutionary essayist of the twentieth century, the Harvard palaeontologist Stephen Jay Gould, was fond of a philosophical thought experiment of his own that he called 'replaying life's tape' (Gould, 1989). That is, consider the history of the evolution of life on Earth to be similar to a video tape of a popular movie. Then imagine what would happen if you could take a copy of the video tape and rewind it to a point early in the movie, erasing everything on the tape that happened after that point, and then rerun the tape to see what happens a second time. Will the historical sequence of events in the evolution of life in the second rerun of the tape resemble the original? Or will evolution take radically different pathways in the second rerun, producing animal and plant forms totally unlike those of the original? Gould argued strongly for the second scenario: 'any replay of the tape would lead evolution down a pathway radically different from the road actually taken . . . the diversity of possible itineraries does demonstrate that eventual results cannot be predicted at the outset. Each step proceeds for cause, but no finale can be specified at the start, and none would ever occur a second time in the same way, because any pathway proceeds through thousands of improbable stages' (Gould, 1989, p. 51).

Gould's (1989) view of the evolution of life as totally unpredictable, totally at the whim of historical contingency, is directly contradicted by the phenomenon of convergent evolution. Let us return to the oft-cited amazing convergence of form seen in the evolution of the cartilaginous fish, the bony fish, the reptiles and the mammals, that we considered in Chapter 3 (Fig. 7.22). In essence, evolution has indeed 'replayed the tape of life' four times in the production of fast-swimming oceanic organisms – and has repeatedly come up with the same morphological solution! The phenomenon of convergent evolution demonstrates to us

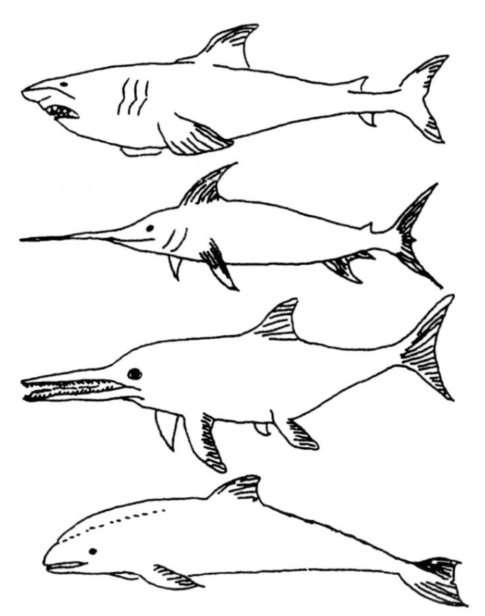

Figure 7.22. Convergent morphologies. Convergently evolved streamlined, fusiform swimming morphologies in the cartilaginous fish (chondrichthyan shark, top), in the bony fish (actinopterygian swordfish, second down), in the reptiles (an ichthyosaur, third down), and in the mammals (a porpoise, fourth down).
Source: Animal sketches redrawn from Funk and Wagnall (1963).

that evolutionary pathways are highly constrained, and thus are, in fact, in principle predictable (Thomas and Reif, 1993). The *degree* to which evolution is predictable is currently under debate, with some proposing a view of predictability in evolution that is radically different from Gould's: the view that the detailed analysis of convergent evolution might one day allow biologists to construct a 'periodic table of life', similar to the 'periodic table of elements' of the chemists. In contrast to western evolutionary thought, the idea that constraint might result in periodic (convergent) biological systems has a long history in Russian science (Popov, 2002).

From a modelling perspective in theoretical morphology, how is the convergent evolution of form shown in Figure 7.22 produced? The cartilaginous, chondrichthyan fish and the bony, actinopterygian fish exhibit a very similar spectrum of morphologies. However, each group does possess some unique morphologies not found in the other, such as manta-ray morphologies in the chondrichthyans and flounder-fish morphologies in the actinopterygians. We can thus spatially

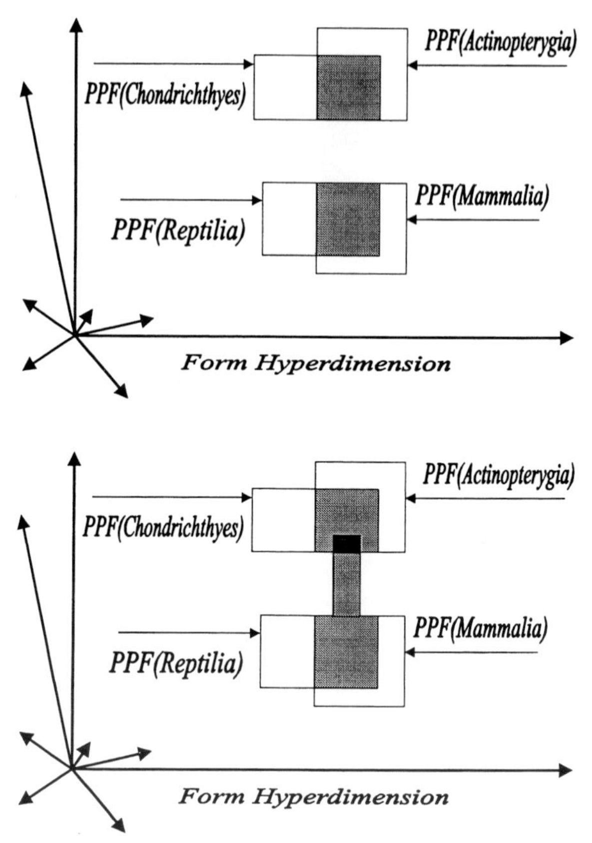

Figure 7.23. Modelling convergence. A spatial representation of a Venn diagram representation of the convergent evolution of form in chondrichthyan fish, actinopterygian fish, reptiles, and mammals, illustrated in Fig. 7.22, in the region of functional possible form (*FPF*). The shaded regions in the upper diagram illustrate the intersection of the fish morphological sets *PPF(Chondrichthyes)* and *PPF(Actinopterygia)*, and the intersection of the tetrapod morphological sets *PPF(Reptilia)* and *PPF(Mammalia)*, but shows no intersection of the four sets. The lower diagram illustrates the convergent evolution of fish-like morphologies by the Reptilia and Mammalia, where the black region of the form hyperspace corresponds to the set intersections *PPF(Chondrichthyes)* ∩ *PPF(Actinopterygia)* ∩ *PPF(Reptilia)* ∩ *PPF(Mammalia)*.

represent the sets of phylogenetically possible morphologies for the two groups as broadly, but not totally, overlapping in form hyperspace (Fig. 7.23). Likewise, we can represent the sets of phylogenetically possible forms available to the reptiles and mammals as broadly, but not totally, overlapping in form hyperspace (Fig. 7.23). In general, however, we consider fish to be fish and land animals to be land animals, and the two groups to be morphologically distinct. That is, we might be tempted to conclude, given their very different ecologies and genetic legacies, that:

$$PPF(Chondrichthyes) \cup PPF(Actinopterygia) \cap PPF(Reptilia) \cup PPF(Mammalia) = \oslash.$$

We know, however, that astonishingly this is not the case (Fig. 7.22). We know that the following intersections actually exist:

$$PPF(Chondrichthyes) \cap PPF(Actinopterygia) \cap PPF(Reptilia) \cap PPF(Mammalia) \neq \oslash.$$

One possible evolutionary pathway to bring about the intersection of these four morphological sets in form hyperspace is illustrated in Figure 7.23. Here the boundary of the set *PPF(Reptilia)* and the boundary of the set *PPF(Mammalia)* are shown to have expanded into the region of form hyperspace occupied by the previously existent set intersection *PPF(Chondrichthyes)* ∩ *PPF(Actinopterygia)*. This hypothesis concludes that first the reptiles, and then the mammals, have separately and independently evolved new genetic coding that allow them to produce the convergent morphologies.

This possibility is not the only one, however. All of the land tetrapods are the descendants of the ancient lobe-finned, sarcopterygian fish (Fig. 7.24). An alternative evolutionary scenario to Figure 7.23 is illustrated in Figure 7.25. Here we begin with the two morphologically similar, but not totally overlapping, fish-form sets *PPF(Actinopterygia)* and *PPF(Sarcopterygia)* in form hyperspace. The evolution of the tetrapods from the sarcopterygian fish (Fig. 7.24) is modelled as the expansion of the boundaries of the phylogenetically possible set of forms originally available to the sarcopterygians to include the region of form hyperspace that contains existent tetrapod morphologies (Fig. 7.25), creating a new morphological set *PPF(Tetrapoda)*. Although later tetrapods actually develop morphologies that only occur in the shaded region of the spatial representation of the set *PPF(Tetrapoda)*, and not the total set *PPF(Tetrapoda)*, the convergent evolution of fish-like morphologies by the reptiles, and then the mammals, is now simply modelled as the reactivation of their ancient genetic legacy from the

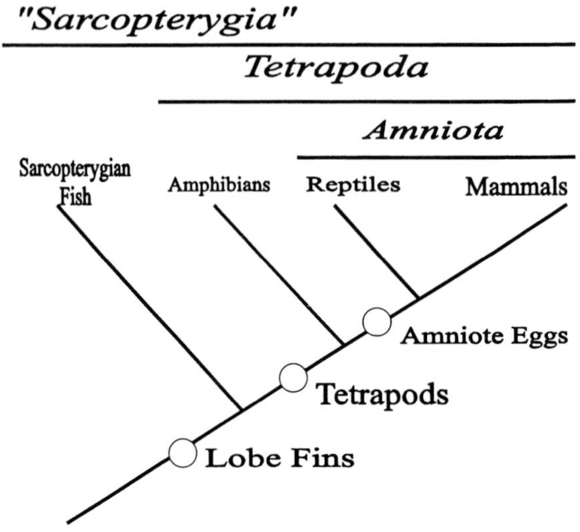

Figure 7.24. A cladogram of the evolutionary relationships of the sarcopterygian fish and the tetrapods.

sarcopterygians, and not the evolution of new genetic coding, as in Figure 7.23. Which model, Figure 7.23 or 7.25, is correct? As before, in our considerations of horse evolution, this is a theoretical question for the empirical geneticists and experimentalists to consider. It may well be that a considerable amount of convergent morphological evolution is the result of convergent developmental mechanisms in organisms that are distantly related (Wray, 2002).

The phenomenon of convergent evolution means that there are a limited number of ways of making a living in nature, a limited number of ways of functioning well in any particular environment. The direction of evolution is highly *constrained*: no matter where you begin your journey on an adaptive landscape, if you evolve to be a active-swimming oceanic animal you will wind up in the same region of the landscape; that is, you will converge on the same morphological solution (Fig. 7.22). Contrary to Gould's (1989) dictum that 'biological evolution has no predictable destination', I predict with absolute confidence that if any large, fast-swimming organisms exist in the oceans of Europa – far away in orbit around Jupiter, swimming under the perpetual ice that covers their world – then they will have streamlined, fusiform bodies; that is, they will look very similar to a porpoise, an ichthyosaur, a swordfish, or a shark (Fig. 7.22).

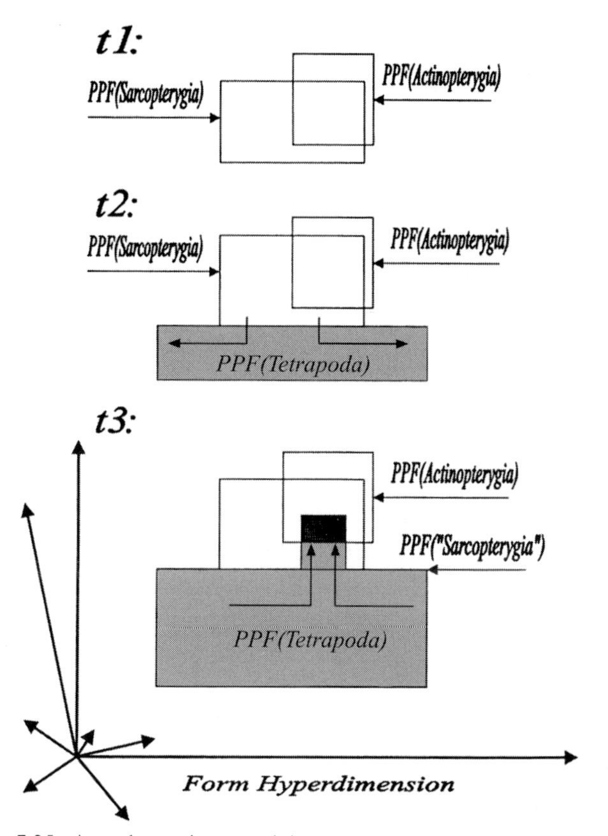

Figure 7.25. An alternative spatial representation of the convergent evolution of fish-like morphologies in the tetrapods
At time *t1* (upper figure) only the morphological sets for the bony fish, *PPF(Actinopterygia)*, and the lobe-finned fish, *PPF(Sarcopterygia)*, are in existence. At time *t2* (middle figure), the evolution of the tetrapods from the lobe-finned fish is represented as an expansion of the boundaries of the original morphological set *PPF(Sarcopterygia)* to create the new set *PPF(Tetrapoda)*. At time *t3* (bottom figure) the convergent evolution of actinopterygian-like morphologies is represented as simply movement within the existent set *PPF(Tetrapoda)*, where the shaded region represents the realized, existent tetrapod morphologies within the total set of phylogenetically possible morphologies for the tetrapods.

In connexion with the present discussion, the existence of such a hypothetical organism would provide direct evidence for the modelling scenario shown in Figure 7.23 as it would represent convergent evolution in an extraterrestrial, Europan, lifeform independent of the origin of life on Earth.

Is the concept of phylogenetic constraint of heuristic value only?

Considering all the possible temporal shifts in the boundaries of the set *PPFx* within the set *FPF* discussed in the previous section, from expansion (Fig. 7.17) to contraction (Fig. 7.18) to shifting (Figs. 7.19 and 7.20) and to fragmentation (Fig. 7.21), one might conclude that anything is possible, given enough evolutionary time. That is, that the concept of phylogenetic constraint is an interesting way of thinking about how the evolutionary process might occur, but that it is impossible to use the concept in the actual analysis of evolution and thus that the concept is of heuristic value only. This question regularly appears in the works of evolutionary biologists, from the simple statement that the constraint concept is 'central to current perceptions of the evolutionary process, but operationally, it is difficult to apply' (Schwenk, 1994, p. 251) to the rhetorical question: 'phylogenetic constraint in evolutionary theory: has it any explanatory power?' (McKitrick, 1993, p. 307). On the other hand, other researchers have extended the concept of phylogenetic constraint to modelling ecological phenomena, and not just morphological (Cattin *et al.*, 2004).

Theoretical morphospace techniques offer the powerful possibility of actually using the concept of phylogenetic constraint as an analytical tool. The discipline of theoretical morphology is in its infancy (McGhee, 1999, 2001a), yet already we have a very possible example of phylogenetic constraint in operation: the inability of the nautilid cephalopods to invade the empty morphospace previously occupied by the ammonoid cephalopods (Fig. 7.16). The morphologies present within the empty region of morphospace clearly belong to the set of functional possible form, *FPF*, a fact the ammonoids clearly demonstrated in their long evolutionary history. These currently nonexistent forms would function just as well today as they did for the ammonoids in the Palaeozoic, why then do they remain nonexistent? Why have not the related swimming cephalopods, the nautilids, evolved these functional forms in the 65 million years following the ammonoids demise? It is difficult to escape the conclusion that phylogenetic constraint is in operation, and that natural selection is unable to overcome this constraint. McKitrick (1993, p. 307) defines phylogenetic constraint to be 'any result or component of the phylogenetic history of a lineage that prevents an anticipated course of evolution in that lineage'. In Figure 7.16 the technique of theoretical morphospace analysis has actually demonstrated *an anticipated course of evolution for a lineage that has been prevented.* Under the expectations of the theory of natural selection, the functional forms present in the empty region of morphospace should have been discovered by the nautilids in the past 65 million years. Yet they have not.

There remains one final constraint that might produce empty regions within functional possible morphospace, that might produce the nonexistence of forms that nevertheless clearly belong to the set intersection $PPFx \cap FPF$ (Fig. 7.5). This is the constraint of development, the final determinant of the actual subset of existent form for lineage x (Fig. 7.6). We shall consider the possible analysis of this challenging intrinsic constraint in the next chapter.

8

Evolutionary development in theoretical morphospace

> The heuristic power of building theoretical morphospaces rests on the capability of generating hypothetical morphologies out of real processes, thus surpassing the usual analytical observation of natural occurrences. At some level, any experimental manipulation involving gain and loss of gene function is a strategy that parallels morphospace building. In both cases, natural occurrences are violated, and new forms appear that have to be explained with normal biological processes. The gain in insight is enormous: looking at the logic of theoretical occurrences can single out the logic of real occurrences.
>
> *Rasskin-Gutman and Izpisúa-Belmonte (2004, p. 411)*

The concept of developmental constraint

In Chapter 7 the concept of *developmental constraint for species x* is defined as the boundary between the two sets $DPFx = \{f|\ f =$ developmentally possible forms for species $x\}$ and $DIFx = \{f|\ f =$ developmentally impossible forms for species $x\}$ (Fig. 7.5). Forms that belong to the set *DPFx*, developmentally possible form for species *x*, must be phylogenetically possible and geometrically possible, belonging to the sets *PPFx* and *GPF*, but can either be functional ($DPFx \cap FPF$) or lethal ($DPFx \cap NPF$). Actual existent form for species *x* can only belong to the former set intersection, such that potential existent form for species *x*, *PEFx*, is constrained to be $PEFx \subset DPFx \subset FPF \subset GPF = \{f|\ f \in GPF, f \in FPF, f \in DPFx\}$ (Fig. 7.6).

A key feature of the developmental constraint concept from a theoretical morphospace perspective is the possibility that there might exist empty regions of morphospace containing potential forms that are phylogenetically possible for species x but that nevertheless cannot be developed by species x, $PPFx \cap DIFx$ (Fig. 7.6). That is, there potentially exists a set of forms, f, for which species x possesses the genetic coding necessary to produce these forms but does not possess the developmental mechanisms necessary for producing these forms, the set $\{f | f \in PPFx, f \notin DPFx\}$.

How tenable is such a concept? Alberch (1982, 1991) has argued that there are two ways to conceptualize the relationship between the genotype and the phenotype. One view holds that genes are the direct determinants of organic form, the phenotype. That is, the genes code for a particular developmental process, which in turn produces the resultant morphology (Fig. 8.1). If this were true, then all existent form could be directly reduced to the DNA coding, the genes, of the organism. In this reductionist view, developmental constraint does not exist as a separate constraint, as development itself is directly reducible to the genotype, thus only phylogenetic constraints exist. That is, the set $\{f | f \in PPFx, f \notin DPFx\} = \oslash$ because $DPFx = PPFx$.

Alberch (1982, 1991) has argued that this reductionist view is false, and that the development of organic form is much more complex and interactive (Fig. 8.2). In this second view, genes code for protein synthesis that either regulates the expression of other genes or regulates the physico-chemical processes of cellular development. Cellular morphogenesis in turn leads to cell-cell interactions in the developing tissue geometry. In particular, Alberch (1991) argued for 'induction events' in the developmental process, when the geometry of tissue development

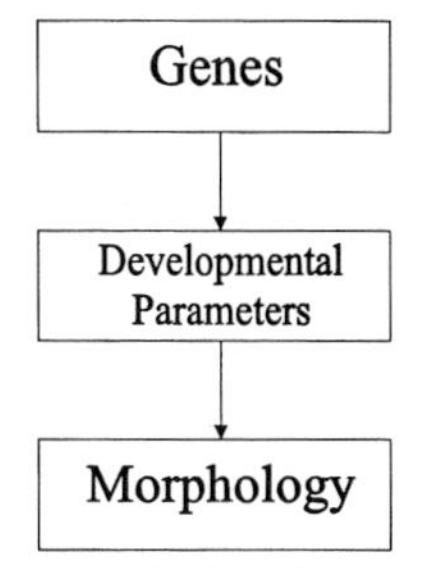

Figure 8.1. A hierarchial, reductionist view of development. Here genes code for a particular developmental process, which in turn produces the resultant morphology.
Source: Modified from Alberch (1991).

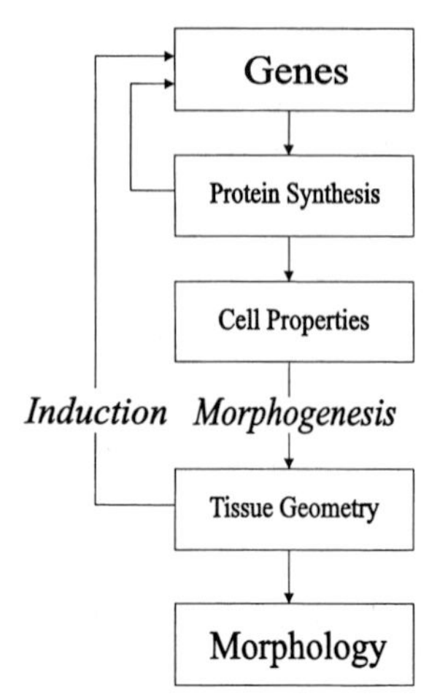

Figure 8.2. An alternative view of development. Here genes are only one part of a feedback cycle of morphogenesis. In this view morphogenesis is not directly reducible to the genotype as it is the result of tissue development itself.
Source: Modified from Alberch (1991).

brings into contact cells with different developmental histories. In such induction events, the expression of genes in some cells may be halted, or turned off, while other genes may be activated, or turned on. Thus a feedback cycle exists in the developmental process, and this cycle is not directly reducible to the genotype as it is the result of tissue development itself (Fig. 8.2). In this alternative, nonreductionist, view it is entirely possible that genes might be present that code for a developmental process that fails at the cellular development or tissue development stage, and thus that the set $\{f \mid f \in PPFx, f \notin DPFx\} \neq \oslash$ and that a separate type of intrinsic constraint, developmental constraint, exists in addition to phylogenetic constraint.

Debate over the concept of developmental constraint sparked several notable conferences in the early 1980s (Bonner, 1982; Maynard Smith *et al.*, 1985) which continue to the present day (see, for example, Müller and Newman, 2003; Callebaut and Rasskin-Gutman, 2005). Keller (2002, p. 3) characterizes the current developmental debate as the calling into question of the genetic form of explanation that

> has come to dominate biological thought over the last few decades – the assumption that a catalogue of genes for an organism's traits will constitute an "understanding" of that organism. Yet, an increasing number of biologists

> are beginning to argue that no such catalogue – not even the sequence of the entire genome – can suffice to explain biological organization. The reason most commonly offered for their skepticism regarding genetic explanations is that the regulatory apparatus for turning genes on and off is distributed throughout the organism.

Indeed, the argument has been made that little, if anything, can be learned from population genetics in the analysis of the developmental changes that occur in the origination of new evolutionary innovations (Müller and Wagner, 1991; Wagner, 2001).

Transformation theory and theoretical morphospaces

An early attempt to use the concept of developmental constraint as an analytical tool can be seen in the *theory of transformations* of D'Arcy Thompson's classic book, *On Growth and Form*. Thompson (1942) was interested in the possible topological pathways by which one organic form might be transformed into another. His analytical method involves superimposing a rectangular cartesian grid over the form of an original species, then measuring the deformations of that grid that are necessary in order to superimpose it over a different, but related, species. By this method he hoped to demonstrate the mathematical relationship of the original grid to the transformed grid, and thus that the comparison 'of the new coordinate system to the old ... will furnish us with some guidance as to the 'law of growth', or play of forces, by which the transformation has been effected' (Thompson, 1942, p. 1048).

An example of Thompson's transformation analysis of development is given in Figure 8.3 for the porcupine-fish *Diodon* and the sunfish *Orthagoriscus*. He argued that the development of a sunfish-type morphology from a more common porcupine-fish-type morphology can be modelled by a series of hyperbolic deformations of the original rectangular cartesian grid superimposed over the porcupine fish (Fig. 8.3). That is, the development of a sunfish-type morphology required the progressive differential expansion of growth gradients in the posterior region of the fish, while growth gradients in the the anterior region of the sunfish remain essentially unchanged from that of the porcupine fish.

A second example of Thompson's differential developmental analyses is given in Figure 8.4 for the skulls of a human, chimpanzee and baboon. Thompson's grid-deformation analyses indicated to him that the chief

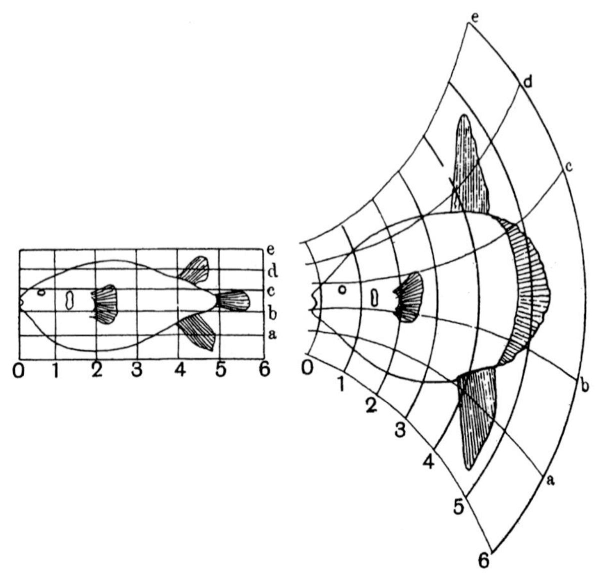

Figure 8.3. Thompson's differential developmental analysis shows morphological transformations necessary in order to turn a porcupine-fish-type morphology (left figure) into a sunfish-type morphology (right figure). Source: From *On Growth and Form*, by D'A. W. Thompson. Copyright © 1942 by Cambridge University Press and reprinted with the permission of the publisher.

developmental differences in the evolution of human-type skulls from related primates lay in the reduction of growth gradients in the snout, or maxillary, region of the skull and the vast expansion of growth gradients in the brain-case region. He considered this developmental transformation to be continuous across the primates, differing only in the degree of deformation of the coordinate grid, with the baboon at one extreme and the human at the other, with the chimpanzee in an intermediate position (Fig. 8.4).

We now know that neither the baboon nor the chimpanzee are in an ancestor-descendant relationship to modern humans, yet we are still all primates and the chimpanzees are genetically very close cousins to us indeed. Thompson's vision was to seek to discover something about the developmental processes of the entire group of primates by his method of transformational analysis.

From the perspective of developmental constraint, what is more important is not the demonstration of the types of developmental transformation that are necessary to produce observed organic forms,

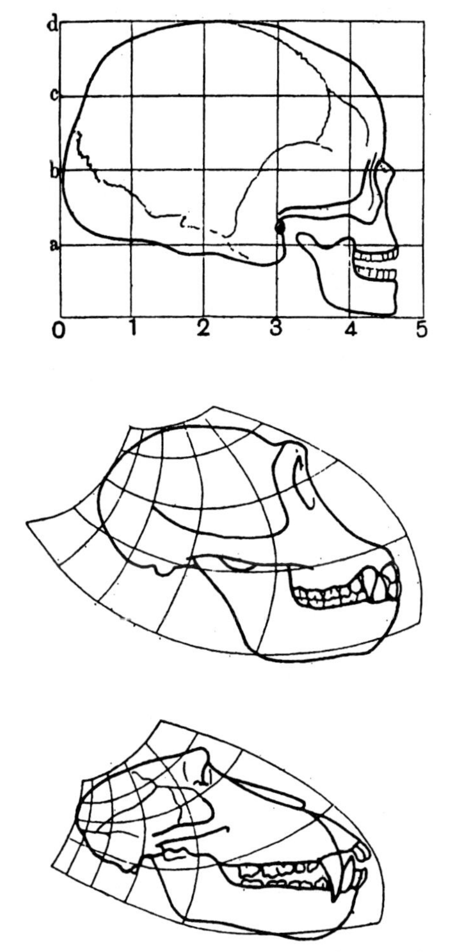

Figure 8.4. Developmental grids showing morphological transformations from a human-type skull (top figure) to a chimpanzee-type skull (middle figure) to a baboon-type skull (bottom figure).
Source: From *On Growth and Form*, by D'A. W. Thompson. Copyright © 1942 by Cambridge University Press and reprinted with the permission of the publisher.

but rather the *limits of developmental transformation*, the demonstration of transformations that are *impossible*. That such limits exist is a key feature of the concept of developmental constraint. Thompson (1942) was active in promoting his analytical technique of developmental transformations, rather than seeking out cases where such transformations are impossible. Yet is is clear that he understood that such limits existed. He considered that he was engaged in the study of that 'Science of Form which deals with the forms assumed by matter under all aspects and conditions, and, in a still wider sense, with forms which are theoretically imaginable' (Thompson, 1942, p. 1026), implying that there exist forms that are theoretically unimaginable or impossible. He specifically stated that his analytical method of developmental transformations was limited to the comparison of organisms that were developmentally similar:

> We should fall into deserved and inevitable confusion if, whether by mathematical or any other method, we attempted to compare organisms separated far apart by Nature and in zoological classification. We are limited, both by our method and by the whole nature of the case, to the comparison of organisms such as are manifestly related to one another and belong to the same zoological class. For it is a grave sophism, in natural history as in logic, to make a transition into another kind.
>
> *(Thompson, 1942, p. 1034)*

This statement does not clearly differentiate the concepts of phylogenetic constraint and developmental constraint, yet Thompson (1942) in these sections of his book was concerned with the analysis of development, rather than phylogeny.

D'Arcy Thompson's ideas on the analysis of developmental transformation has inspired generations of morphometricians, scientists concerned with the precise measurement of form in individual organisms and the precise comparison of those measurements among different individuals (see Bookstein, 1977, 1997). However, it is only recently that D'Arcy Thompson's ideas have been adapted to theoretical morphospace analysis (Rasskin-Gutman and Buscalioni, 2001; Rasskin-Gutman, 2003).

Thompson was interested in the relationships of dinosaurs and birds, and Figure 8.5 illustrates his transformational analysis of the evolutionary development of pelvic structures in ancient Jurassic and Cretaceous toothed birds. Both of the pelvic structures shown in Figure 8.5 actually exist – but can the analytical techniques of Thompson be extended to *nonexistent forms*, to reveal developmental transformations that do not exist? The answer is yes, as Rasskin-Gutman and Buscalioni (2001) have demonstrated. A full range of both existent and nonexistent pelvic structures in theropod dinosaurs (which includes the birds) is illustrated in Figure 8.6, which demonstrates the ability of the analytical techniques of theoretical morphology actually to create those forms that are 'theoretically imaginable' (Thompson, 1942, p. 1026).

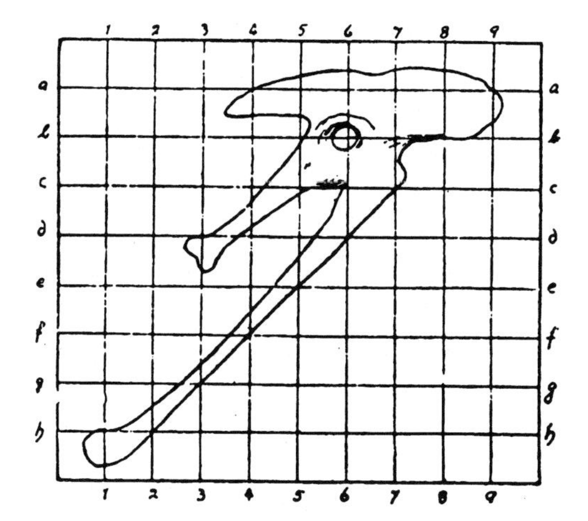

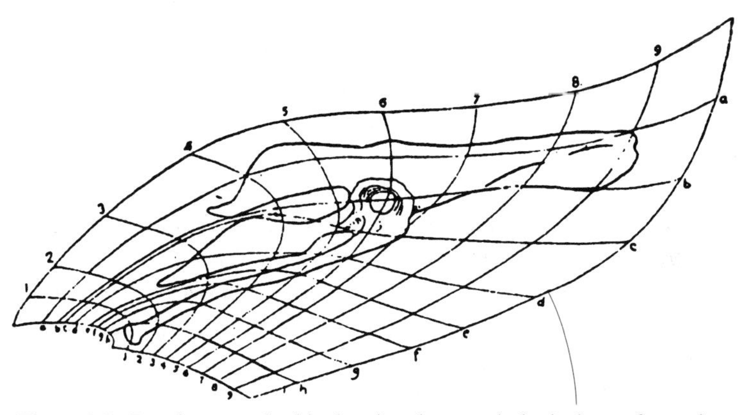

Figure 8.5. Developmental grids showing the morphological transformation from the pelvic structure of an ancient Jurassic bird (*Archaeopteryx*, top figure) to a more derived Cretaceous bird (*Apatornis*, bottom figure). Source: From *On Growth and Form*, by D'A. W. Thompson. Copyright © 1942 by Cambridge University Press and reprinted with the permission of the publisher.

Epigenetic landscapes and theoretical morphospaces

Figure 8.6 illustrates the conceptual extrapolation of D'Arcy Thompson's theory of developmental transformations into the realm of theoretical morphospace. All of the morphologies shown are mathematical distortions, 'developmental transformations' in the sense of Thompson

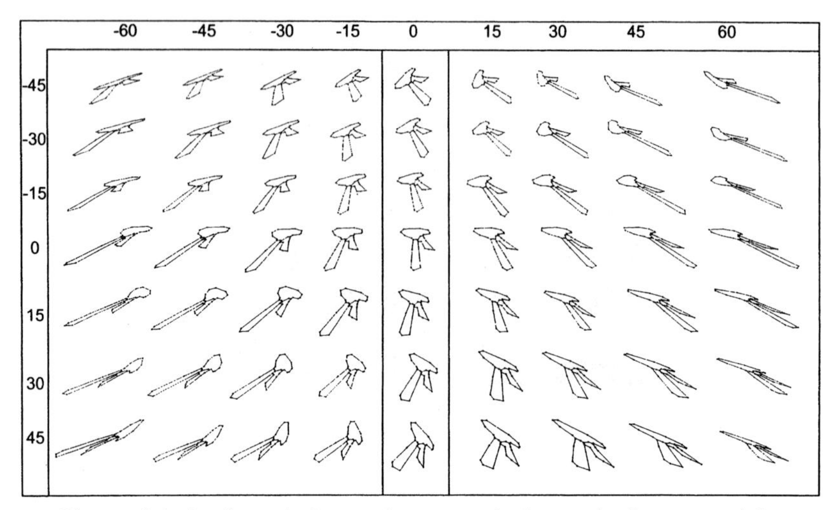

Figure 8.6. A theoretical morphospace of theropod dinosaur pelvic structures.
Source: From Rasskin-Gutman and Buscalioni (2001). Copyright © 2001 by the Paleontological Society and reprinted with permission.

(1942), of the actual pelvis of the theoropod dinosaur *Deinonychus antirhopus* (Rasskin-Gutman and Buscalioni, 2001). The argument could thus be made that all of the hypothetical forms shown in Figure 8.6 were (or are, if one includes the birds) actually developmentally possible for the theropod dinosaurs. Or are they?

The concept of mapping developmentally possible and developmentally impossible regions of morphospace was first proposed by Waddington (1957, 1975) in his classic graphical portrayals of *epigenetic landscapes*. Unlike the adaptive landscapes that we have considered thus far, where morphological coordinates in the flat plain of the landscape represents nonfuctional morphologies and various degrees of functional morphologies are found on the slopes and peaks of adaptive hills, the plain of an epigenetic landscape represents morphological coordinates that are developmentally impossible (Fig. 8.7). There are no topographic peaks or mountains in Waddington's epigenetic landscapes – in contrast, the flat plain of the landscape is cut by a series of winding and branching depressions that cut through it, very similar to the pattern produced by river valleys in an actual geographical landscape (Fig. 8.7).

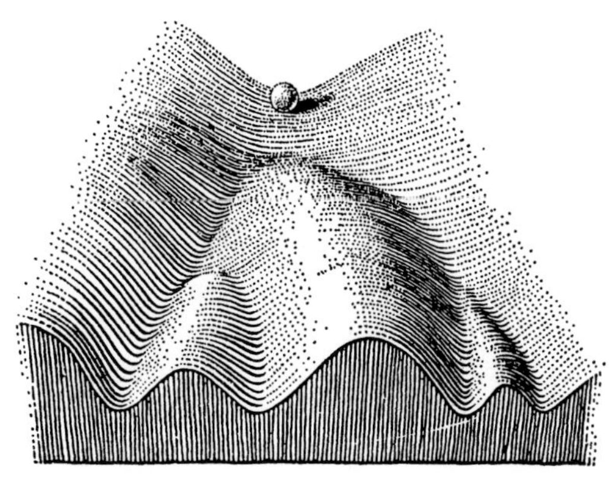

Figure 8.7. The concept of the epigenetic landscape. The developing embryo is represented by the ball, which may roll down several different developmental valleys, each of which results in a different morphological result.
Source: From "A catastrophe theory of evolution", by C. H. Waddington, *Annals of the New York Academy of Sciences*, volume 231, pp. 32–42. Copyright © 1974 by the New York Academy of Sciences, U.S.A., and reprinted with permission.

In an epigenetic landscape, only the morphological coordinates that occur in the river valleys and streams are developmentally possible. Wright (1932) envisaged evolution as a process where populations of individuals climb adaptive peaks under the influence of natural selection. Waddington (1957, 1975) envisaged development as a process where a single individual moves along the bottom of a depression in an epigenetic landscape, similar to the process of rolling a ball down a valley (Fig. 8.7). Along the way, in the process of the development of the individual, the ball might encounter branchings in the landscape, alternate developmental pathways that lead off to different regions in the epigenetic landscape. Once the ball enters a particular valley, it is committed to a particular developmental pathway. This progressive channeling of development down a limited number of pathways, towards a limited number of possible morphological outcomes, was termed *developmental canalization* by Waddington (1957, 1975).

On the other hand, Waddington did think that it might be possible, in certain natural selection circumstances, for a developing individual to jump over the watershed, the flat plain of the epigenetic landscape, from one developmental valley to another alternative developmental pathway. The concept that such abrupt developmental jumps might be possible he termed the *catastrophe theory of evolution*, and the watershed between two alternative developmental valleys he designated as a *catastrophe surface* to be jumped in the epigenetic landscape. The concept of a sudden jump from one region to another in a continuous topology intrigued the mathematician René Thom, who devoted an entire book, *Structural Stability and Morphogenesis*, to exploring the mathematical properties of various topological foldings of space that would permit such jumps to occur. Thom's (1975) topological models subsequently became known as *catastrophe theory* in developmental biology. Pere Alberch's early experimental work in amphibian development (Alberch and Gale, 1985; Oster *et al.*, 1988) remains a classic argument in favour of the idea that gaps or discontinuities might exist in epigenetic landscapes, due to the internal constraints of development. More recent experimental and theoretical work at the molecular level (Stadler *et al.*, 2001) and modelling (Stadler, 2002; Rasskin-Gutman, 2005) continue to argue for the concept of developmental discontinuities in adaptive landscapes, topological arguments similar to Waddington's ideas of developmental jumps in epigenetic landscapes.

The main conceptual contribution of Waddington's epigenetic landscapes is the visualization of development as the movement of the embryo across a continuous surface along restricted pathways: 'I have urged that we should think not in terms of homeostasis, but rather of homeorhesis, the stabilization not of a stationary state, but of a pathway of change in time. Instead of picturing the resistance-to-change of a phenotype as a minimum in a space that is multidimensional for physiochemical variables but does not include time, I envisage it as a set of branching valleys in a multidimensional space that includes a time dimension, along which the values extend' (Waddington, 1975, p. 258). The actual movement of a given individual across the epigenetic landscape, along its particular set of developmental pathways, thus can be visualized as the *ontogenetic trajectory* for that individual (Olsen and Miller, 1958). At each given point in time, t, the developing organism has a particular set of coordinates that specify its phenotypic position, p, on the epigenetic landscape, and its total ontogenetic trajectory is the function dp/dt (Alberch *et al.*, 1979; Rice, 1997).

Such a spatial representation of the developmental process is immediatly transferable to the concept of theoretical morphospace (Richtsmeier and Lele, 1993). Indeed, some recent studies in developmental biology refer to developmental pathways within a morphospace rather than a epigenetic landscape (see, for example, Figure 16.1A of Streidter, 2003).

Analysing development in theoretical morphospace

If organisms grow anisometrically then they, by definition, experience changes in morphology as they grow. In a theoretical morphospace context, this means the morphological coordinates of juvenile organisms within the morphospace may be quite different from the morphological coordinates of an adult; in essence, the organism moves from one region of the morphospace to another during development. Thus development will produce ontogenetic trajectories of morphological coordinate change in theoretical morphospace, analogous to ontogenetic trajectories in an epigenetic landscape. In the special case of isometric growth, such ontogenetic trajectories will collapse to a single point, as the developing organism does not move within the morphospace with growth.

Although Dave Raup's early models of morphogenesis were isometric, he did explore the consequences of anisometric growth in theoretical morphospace. In Figure 8.8 is given a cross-section of an ammonoid shell that exhibits strong anisometric growth. This extinct species, *Paracravenoceras ozarkense*, started out life with a shell that is advolute, where the whorls do not overlap one another. During growth, however, the shell became progressively more involute, such that in the adult shell the outermost whorl almost completely overlaps the younger whorls (Fig. 8.8). Measurements taken from this species, and plotted in Raup's theoretical morphospace of ammonoid form (see discussion in Chapter 4), are given in Figure 8.9. The plot of the ontogenetic trajectory of *P. ozarkense* reveals that most of the developmental change in its shell form occurred in the *D*-dimension of the morphospace: the animal started out life with a shell with a *D*-value of around 0.4, moved briefly to a slightly higher *D* of 0.5, then progressively moved through the morphospace to lower and lower values of *D*, finally achieving adulthood with a shellform with a *D*-value of almost 0.1 (Fig. 8.9). Interestingly, Raup (1967) discovered that none of the ammonoids he examined, even one with such marked anisometric growth as *P. ozarkense*, exhibited ontogenetic trajectories that ventured outside the frequency distribution of form exhibited by his total sample of ammonoids (Fig. 4.7). Morphological development in the ammonoids was apparently limited by the functional need to remain in the vicinity of the ridge of ammonoid adaptive peaks (Fig. 5.1).

Another example of the actual analysis of morphological development in theoretical morphospace is given in Figure 8.10. Illustrated is

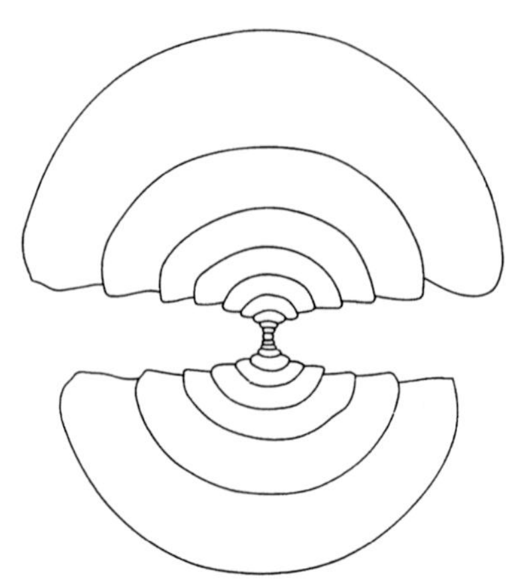

Figure 8.8. Ammonoid development. A cross-section through the shell of the Carboniferous ammonoid *Paracravenoceras ozarkense*, showing marked changes in shell cross-sectional morphology from early growth stages (whorls in the centre of the shell) to the adult (outermost whorl).
Source: From Raup (1967). Copyright © 1967 by SEPM (Society for Sedimentary Geology) and reprinted with permission.

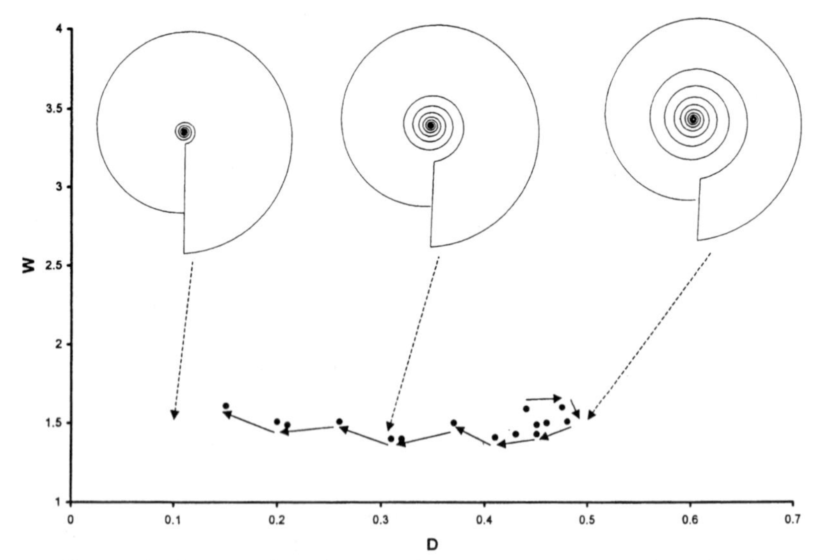

Figure 8.9. Ontogenetic trajectory of *Paracravenoceras ozarkense* in the theoretical morphospace of ammonoid form discussed in Chapters 4 and 5. Computer simulations show typical isometric adult shell-forms located in three regions of the morphospace.
Source: Ontogenetic data from Raup (1967).

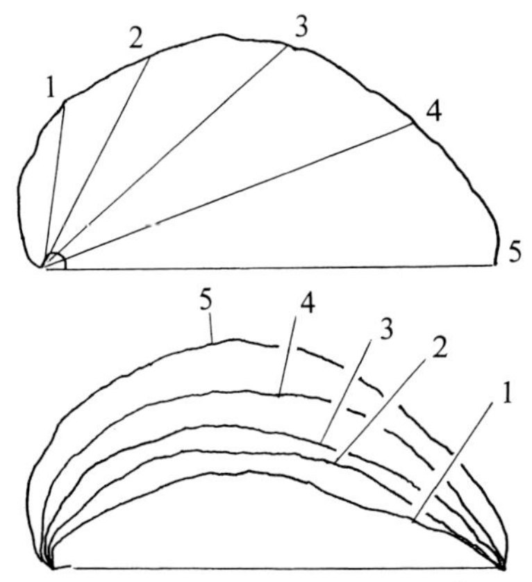

Figure 8.10. Brachiopod development. Upper figure: a cross-section through the dorsal valve of the shell of the Ordovician brachiopod *Lycophoria nucella*, where five growth stages are marked at the valve margin. Lower figure: each of the five growth stages marked above have now been expanded and scaled such that they all have the same length, thus comparatively illustrating the marked developmental changes that occurred in shell convexity during the life of the animal.
Source: Modified from McGhee (2001b).

a cross-section through the dorsal valve of the shell of the extinct brachiopod *Lycophoria nucella*, a species that exhibits strong anisometric growth. The animal started out life with a dorsal valve that is only slightly convex, but during growth the valve progressively became more convex, such that the adult valve is almost hemispherical (Fig. 8.10). Measurements taken from both valves of the shell of this species, and plotted in the theoretical morphospace of brachiopod form (see discussion in Chapter 5), are given in Figure 8.11. The plot of the ontogenetic trajectory of *L. nucella* reveals marked developmental changes in both the dorsal and ventral valves of the shell: the animal started out life with a shell with W-values of around 10^7 in both valves, having a very flattish shell with a large surface area and small internal volume. As the animal progressively became larger with growth, however, it moved through the morphospace to regions with lower and lower values of W in both valves, such that expansion in the length of the shell virtually ceased in the adult animal ($W = 0$). The adult animal has a highly convex, roughly spherical shell with a very small surface area and a very large internal volume.

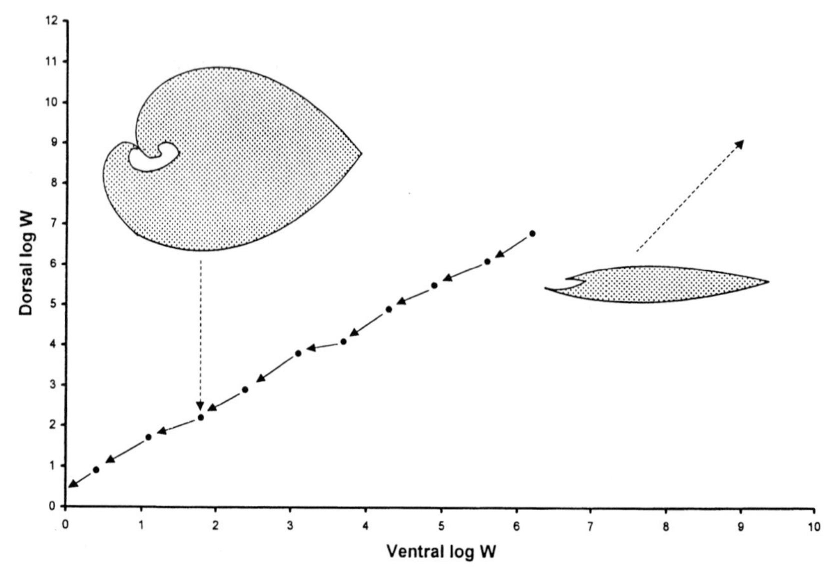

Figure 8.11. Ontogenetic trajectory of an individual of the species *Lycophoria nucella* through the theoretical morphospace of brachiopod shell form (McGhee 1980a). Computer simulations show typical isometric adult shell-forms located in two regions of the morphospace.
Source: Ontogenetic data from McGhee (1980a).

Note that, in contrast to the ammonoids (Fig. 8.9), the adult morphology of *L. nucella* had an ontogenetic trajectory that did move outside of the frequency distribution of form exhibited by the total sample of brachiopods (Fig. 5.4). Indeed, the trajectory crosses into the forbidden area of the morphospace, seemingly penetrating the geometric constraint barrier at W equal to 10^2, imposed by articulation constraints (see discussion in Chapter 5). Many brachiopods thus found a way around the geometric constraint of articulation limitations, imposed by the necessity of avoiding whorl overlap in both valves, by growing anisometrically. They began life with high W values, far from the region of the whorl-overlap geometric-constraint boundary at $W = 10^2$ (Figs. 5.3 and 5.4), then progressively decreased the magnitude of W with growth (Fig. 8.11). Their juvenile shells thus are flattish in the critical posterior region of the shell where the articulation must be maintained, but the shell then becomes more and more convex with growth to culminate with an adult shell that is almost spherical in shape. The much larger adults thus had shells with large internal volumes, which allowed them to house a large lophophore with which to feed.

Subsequent researchers have analysed development in echinoids (Ellers, 1993) and in gastropods (Stone, 1998) by plotting ontogenetic trajectories of these organisms in theoretical morphospace. The field of theoretical morphology is in its infancy, and the number of such studies at present are few. The crucial point is, however, that ontogenetic trajectories and developmental trajectories are not just heuristic concepts. The techniques of theoretical morphospace analysis allow the actual mapping of ontogenetic trajectories for real organisms.

The eventual actual demonstration of developmental constraint boundaries for a group of organisms within theoretical morphospace may come from the analysis of the set of ontogenetic trajectories exhibited by that group of organisms in the morphospace. Are there regions of morphospace that are never crossed by an ontogenetic trajectory? What are the developmental differences, the sequence of morphological changes, that are seen along actual existent ontogenetic trajectories versus nonexistent ontogenetic trajectories in morphospace? For the analyst, plotting growth-stage data for organisms within theoretical morphospaces can reveal the actual ontogenetic trajectories within that morphospace that nature has produced and, by comparison with the empty regions of the morphospace, reveal ontogenetic trajectories – developmental pathways – that are theoretically possible but that have never evolved. Analysis of these nonexistent developmental pathways, similar to the analysis of nonexistent form, can be a powerful tool in the analysis of developmental constraint. Analyses of this type may one day allow us to map the boundaries of developmental constraint for a group of organisms in theoretical morphospace (Fig. 7.5), just as we can map the boundaries of geometric, functional and phylogenetic constraint for real organisms, as demonstrated in Chapter 7.

Theoretical developmental morphospaces?

The dimensional parameters of a theoretical morphospace may be taken either from a geometric model of form, or from a mathematical model of the morphogenesis of form, as discussed in Chapter 4 (Fig. 4.3). However, in the analysis of development, there are many who argue that only the latter approach should be used; that is, the dimensions of the morphospace should be explicitly morphogenetic (Gärdenfors, 2000; Wolfram, 2002).

Other researchers have proposed that a different type of theoretical morphospace, a *developmental morphospace*, be created for the analysis of evolutionary development. Rather than creating a hypothetical spectrum of form, as in a conventional theoretical morphospace, it has been proposed that a theoretical developmental morphospace would create a hypothetical spectrum of developmental possibilities. Eble (2003, pp. 38–39) has argued that the idea of a developmental morphospace

> can be seen as a logical morphological extension of Waddington's metaphor of the "epigenetic landscape" (Waddington, 1957), Goodwin's notion of "epigenetic space" (Goodwin, 1963), and Alberch's rendition of "parameter space" (Alberch, 1989). While such developmental spaces are suitable for probing the genotype-phenotype map (Wagner and Altenberg, 1996), developmental morphospaces are more useful for the inference ... of more focally phenotypic phenomena such as heterochrony, heterotopy, and developmental constraints.

How might such a theoretical 'developmental' morphospace be constructed? For the dimensional parameters of such a morphospace, Rasskin-Gutman and Izpisúa-Belmonte (2004, p. 405) have proposed that 'ontogenetic trajectories can be used as the generative functions that build morphospaces ... theoretical morphospaces can be built using ranges of variations for the parameters of the function that describes the ontogenetic trajectory'. Using the Rasskin-Gutman and Izpisúa-Belmonte (2004) proposal, let see if we can construct a theoretical developmental morphospace for brachiopods. The morphospace containing the ontogenetic trajectory for the brachiopod *L. nucella* (Fig. 8.11) has the dimensional parameters dorsal and ventral W, the whorl expansion rate. The whorl expansion rate is a geometric parameter from a mathematical model of form devised by Dave Raup in his early explorations of theoretical morphospace, as discussed in Chapter 4. This geometric parameter was also used in creating the theoretical morphospace of hypothetical brachiopod forms illustrated in Figure 5.3, within which the ontogenetic trajectory of *L. nucella* is plotted (Fig. 8.11). All of the hypothetical brachiopod shells shown in Figure 5.3 are isometric, W does not change during their growth and they all exhibit no curvature changes in their valves during development. The shell of *L. nucella*, however, does exhibit strong changes in the curvature of its valves with growth (Fig. 8.10), hence the W value of the valves changes with development, resulting in the ontogenetic trajectory within the morphospace seen in Figure 8.11.

In the previous discussion of ontogenetic trajectories within epigenetic landscapes, we saw that an ontogenetic trajectory is the function dp/dt, the change in the phenotypic coordinates of the developing organism in the landscape with time. In the case of *L. nucella*, this ontogenetic trajectory function is $dW/d\varphi$, where time is measured in terms of the angular growth of the spiral of the valve, φ (Fig. 4.4), and the phenotypic trait of the valve is W. In the case of isometric growth, $dW/d\varphi$ is equal to zero, and the convexities of the valves in the brachiopod shell do not change with development. This leaves two developmental possibilities for anisometric growth, either the value of $dW/d\varphi$ is positive, or it is negative. If the value of $dW/d\varphi$ is *positive*, then the convexity of the valve in the brachiopod shell *decreases with growth* (Fig. 8.12). If the value of $dW/d\varphi$ is *negative*, then the convexity of the valve in the brachiopod shell *increases with growth* (Fig. 8.12). We can combine these developmental possibilites in a logical contingency table, thereby creating a theoretical developmental morphospace for

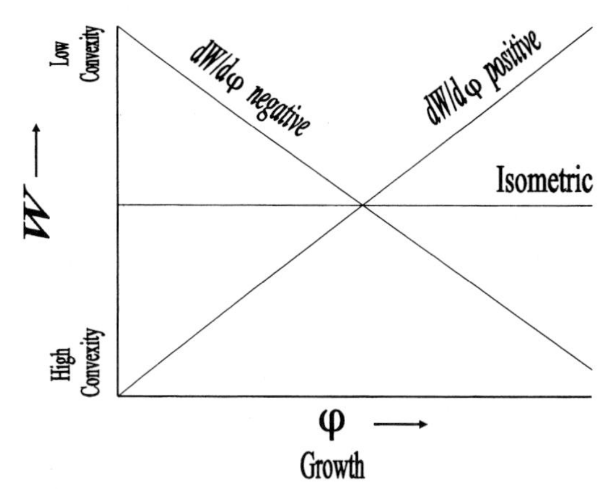

Figure 8.12. Ontogenetic trajectory functions, $dW/d\varphi$, for brachiopod and mollusc shell simulations. The convexity of the shell is inversely proportional to the magnitude of the whorl expansion rate, W (see Fig. 5.3) and growth is measured in terms of the angular length of the spiral, φ (see Fig. 4.4). If growth is isometric ($dW/d\varphi$ equal to zero) no change in the convexity of the shell occurs with development.

brachiopod shells, in which the two dimensions of the morphospace are the ontogenetic trajectories $dW/d\varphi$ for the dorsal and ventral valves of the shell (Table 8.1).

In this developmental morphospace, the central cell contains the developmental possibility (dorsal isometric, ventral isometric) seen in the valves of the computer simulated brachiopods given in Figure 5.3. Computer simulation of the other developmental possibilities would not be that difficult. For example, Figure 8.13 illustrates a computer simulation of a gastropod with strong anisometric growth and an ontogenetic trajectory, $dW/d\varphi$, that is negative.

The creation of such theoretical developmental morphospaces might indeed give us new insights into the processes and limitations of development. Do all the developmental possibilites shown in Table 8.1 actually exist in real brachiopods? We actually have an answer to that question: the empirical analysis of a large sample of brachiopod species reveals that the overwhelming majority of those species with anisometric growth have ontogenetic trajectories that are negative in both valves of the shell (McGhee, 1980a). That is, most brachiopods use the developmental possibility in the upper-left-corner cell of the morphospace (dorsal convexity increase, ventral convexity increase). A much smaller number use the developmental possibility in the middle-left-column cell

Table 8.1. *A theoretical developmental morphospace for hypothetical brachiopod shells proposed here*

VENTRAL VALVE	DORSAL VALVE		
	Negative $dW/d\varphi$	Zero $dW/d\varphi$	Positive $dW/d\varphi$
Negative $dW/d\varphi$	Dorsal Convexity increase, Ventral Convexity increase	Dorsal Isometric, Ventral Convexity increase	Dorsal Convexity decrease, Ventral Convexity increase
Zero $dW/d\varphi$	Dorsal Convexity increase, Ventral Isometric	Dorsal Isometric, Ventral Isometric	Dorsal Convexity decrease, Ventral Isometric
Positive $dW/d\varphi$	Dorsal Convexity increase, Ventral Convexity decrease	Dorsal Isometric, Ventral Convexity decrease	Dorsal Convexity decrease, Ventral Convexity decrease

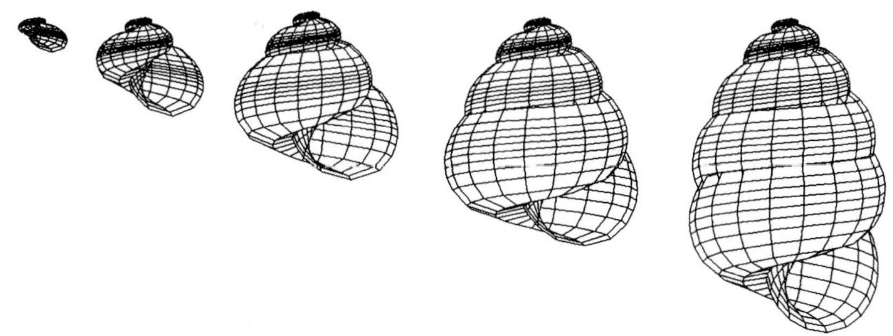

Figure 8.13. Computer simulation of anisometric growth in the gastropod *Gulella*. The ontogenetic trajectory, $dW/d\varphi$, for this organism is negative, such that the curvature of the shell increases with development.
Source: From Cortie (1989), artwork courtesy of M. B. Cortie. Copyright © 1989 by the *South African Journal of Science* and reprinted with the permission of the publisher.

(dorsal convexity increase, ventral isometric), and an even smaller number use the developmental possibility in the top-middle-column cell (dorsal isometric, ventral convexity increase).

No brachiopods examined use any of the developmental possibilities in the other five cells of the morphospace (Table 8.1). That is, a positive ontogenetic trajectory in either of the two valves of the brachiopod shell is not found in the sample of brachiopods examined by McGhee (1980a). Why? In traditional morphospace analyses, we seek to simulate the spectrum of possible form, to reveal which hypothetical forms actually exist in nature and which do not, and then to analyse the causes of nonexistent form. In the case of the developmental morphospace, we seek to simulate the spectrum of possible development, but otherwise the analytical procedure is similar. And, after all, development produces form. The computer simulation of the forms produced by the nonexistent developmental possibilities in Table 8.1 may reveal why they are not used by real brachiopods. It may be simply due to functional constraint. All of the nonexistent developmental possibilites would produce shells that actually decrease their internal volumes, and increase their external surface areas, during growth compared with shells that have the same initial morphological parameters and that grow isometrically (Fig. 5.3). And, as discussed in Chapter 5, brachiopods generally require shells with large internal volumes and small external surface areas for sound biological reasons.

Alternatively, the nonexistence of the five developmental possibilities may be due to developmental constraint. That is, the *developmental processes* look possible, as outlined in the spectrum of possibilites given in Table 8.1, but they in fact *cannot produce* actual form. Computer simulations, using the nonexistent ontogenetic trajectories as model parameters, can reveal whether these developmental processes in fact can produce hypothetical form or not.

At present, the concept of a theoretical *developmental* morphospace analysis of developmental constraint remains just that, a concept. Yet the analytical steps necessary for the conduct of such an analysis are all possible, and it is only a matter of time until someone attempts them.

How does an organism come to be?

In her thoughtful book, *Making Sense of Life: Explaining Biological Development with Models, Metaphors, and Machines*, Evelyn Fox Keller asks the rhetorical question 'How does an organism come to be?' and she proceeds to examine all the myriad ways biologists have attempted to construct explanations of biological development in individual organisms. Specifically, with regards to the work of D'Arcy Thompson, she asks 'What is the basis of his enduring reknown? How are we to account for the high regard in which his celebration of the importance of physical and mathematical models in biology has been held, when the actual use of such models in the development of modern biology has in fact been so minimal?' (Keller, 2002, pp. 52–53). She notes that, although the biologist G. Evelyn Hutchinson once described Thompson's theory of transformations as 'a floating mathematics for morphology, unanchored for the time being to physical science' (quoted in Gould, 1976), the modern discipline of theoretical morphology and the advent of the digital computer have advanced the quantitative study of morphology and might provide the missing anchoring. But, in line with her original observation concerning the work of D'Arcy Thompson, she goes on to note that 'over the last twenty-five years, a number of workers in theoretical morphology have continued to make use of Thompson's framework. Yet … such efforts continue to remain on the margins of contemporary biological research' (Keller, 2002, p. 69).

The thesis of this chapter, and indeed of the entire book, is that the power of the techniques of theoretical morphospace analysis remains largely unrecognized by the biological community. Most evolutionary biologists still think of the adaptive landscape and the epigenetic landscape as heuristic concepts, interesting ways of thinking about life but unusable in the actual analysis of life. The challenge for present and future theoretical morphologists is to dispel this misconception, and to demonstrate to the wider biological community that adaptive landscapes and epigenetic landscapes can be used as actual analytical tools through the creation of theoretical morphospaces. The magnitude of this challenge is the subject of the next, and final, chapter of this book.

9

There is much to be done ...

> A second reason for giving theoretical morphology a good run for its money is that, in many ways, it is unlike other sciences. In *A New Kind of Science* (2002) Stephen Wolfram has controversially suggested that science in the future will be more concerned with algorithms than laws. One wonders how this could be true of all science, but if there is any science of which this is clearly true, that science is theoretical morphology. One of the great success stories of this discipline is the discovery of algorithms that accurately chart the progress of both growth and evolution.
>
> *Maclaurin (2003, p. 465)*

Adaptive landscapes and theoretical morphospaces

The main goal of this book has been to demonstrate to the reader that the concept of the adaptive landscape need not be thought of as merely a heuristic device to conceptualize the process of evolution. The adaptive landscape concept can be put into actual analytical practice through the usage of theoretical morphospaces.

The concept of nonexistent morphology is implicit in Sewall Wright's idea of an adaptive landscape, as is the concept of functional constraint. Theoretical morphospaces allow us to go beyond merely thinking about the possibility of nonexistent morphologies to actually creating, via computer simulation, morphologies and geometries that have never been evolved by life on Earth. Theoretical morphospaces allow us to map the boundaries of functional constraint for a group of organisms within that morphospace.

In contrast to an adaptive landscape, where the dimension of fitness or degree of adaptation is a fundamental feature of the concept, theoretical morphospaces are constructed without any assumptions concerning the adaptive significance of the hypothetical morphologies produced within that morphospace. The absence of actual organic form in that morphospace does not necessarily mean that the hypothetically possible, but naturally nonexistent, morphologies are nonadaptive – something that would be automatically assumed in an adaptive landscape. Thus theoretical morphospaces allow us to go beyond the adaptive landscape concept of functional constraint, and to consider empty regions of morphospace that might be due to geometric, phylogenetic and developmental constraints.

Current progress in theoretical morphospace analyses

The discipline of theoretical morphology is still in its infancy (McGhee, 1999, 2001a). Even so, a considerable number of different forms of life have been the subject of theoretical morphospace analyses to the present date (Table 9.1). Much work has been done with organisms that possess shells, such as brachiopods, cephalopods, gastropods and bivalves. This is due to the fact that all of these organisms grow by simple accretion, and accretionary growth systems are some of the easiest to model mathematically (for a detailed example of such modelling, see McGhee 1999). Other organisms appear to be radically different but in fact can be modelled by geometries that are strikingly similar. These morphologies include those possessed by trees, which are land-dwelling plants, and bryozoans, which are ocean-dwelling animals! Both groups of organisms use branching growth systems, and models designed to produce hypothetical tree morphologies often can be used to model many bryozoan forms as well.

Other groups of organisms, such as arthropods and vertebrates, have very intricate skeletal morphologies and are much more difficult to model. Here theoretical morphologists have begun to analyse these organisms by simply modelling parts of arthropod or vertebrate morphology, rather than the much more complex total.

Most theoretical morphologic simulations these days are programmed in BASIC or C/C++ for the microcomputer, your standard PC. My first simulations of shell form in brachiopods, back in the 1970s, were written in FORTRAN for mainframe computers. Even further

Table 9.1. *Selected groups of organisms that have been the subject of theoretical morphospace analyses. For a detailed review of many of these studies see McGhee (1999)*

MARINE UNICELLULAR ORGANISMS

Silicoflagellates (delicate silica-lattice skeletal forms): *McCartney and Loper* (1989, 1992).

Foraminiferids (delicate calcareous-sphere skeletal forms): *Berger* (1969), *Brasier* (1980), *Signes et al.* (1993), *Tyszka and Topa* (2005), *Tyszka* (2006).

MARINE ANIMALS

Stromatoporoids (ancient forms of calcareous sponges): *Kershaw and Riding* (1978), *Swan and Kershaw* (1994).

Bryozoans (delicate colony forms of moss animals): *McKinney and Raup* (1982), *Cheetham and Hayek* (1983), *McGhee and McKinney* (2000, 2002), *Starcher and McGhee* (2000, 2002), *McKinney and McGhee* (2003, 2004), *McGhee and Starcher* (2006).

Brachiopods (shell forms of the lampshell animals): *Raup* (1966), *McGhee* (1980a, 1980b, 1995, 1999), *Okamoto* (1988), *Ackerly* (1989).

Cephalopods (chambered shell forms of swimming molluscs): *Raup* (1966, 1967), *Chamberlain* (1981), *Ward* (1980), *Bayer and McGhee* (1984), *Saunders and Swan* (1984), *Okamoto* (1988), *Ackerly* (1989), *Dommergues, Laurin, and Meister* (1996), *Korn* (2000), *Checa, Okamoto, and Keupp* (2002), *Wolfram* (2002), *Saunders, Work, and Nikolaeva* (2004), *McGowan* (2004), *Hammer and Bucher* (2005).

Gastropods (spired shell forms of the snails): *Raup and Michelson* (1965), *Raup* (1966), *Davoli and Russo* (1974), *Kohn and Riggs* (1975), *Rex and Boss* (1976), *Cain* (1977), *Williamson* (1981), *Okamoto* (1988), *Ackerly* (1989), *Schindel* (1990), *Stone* (1996, 1998, 1999, 2002, 2004), *Wolfram* (2002).

Bivalves (bivalved shell forms of clams, scallops, and kin): *Raup* (1966), *Savazzi* (1987), *Okamoto* (1988), *Ackerly* (1989, 1992), *Wolfram* (2002), *Ubukata* (2000, 2001, 2003a, 2003b, 2005).

Echinoderms (plated skeleton forms of echinoids and kin): *Waters* (1977), *Ellers* (1993), *Kendrick* (2007).

Hemichordates (delicate colony forms of graptolites): *Fortey* (1983), *Starcher and McGhee* (2003), *McGhee and Starcher* (2006).

Urochordates (larval swimming morphologies): *McHenry and Patek* (2004).

Chondrichthyans (denticle scale skins of sharks): *Reif* (1980).

LAND PLANTS

From primitive stem plants to bushes, shrubs, and trees: *Honda and Fisher* (1978), *Niklas and Kerchner* (1984), *Ellison and Niklas* (1988), *Niklas* (1986, 1997a, 1997b, 2004, 2006).

Leaf morphologies: *Wolfram* (2002), *Zwieniecki, Boyce, and Holbrook* (2004).

LAND ANIMALS

Reptiles (archosaur pelvic morphologies): *Rasskin-Gutman and Buscalioni* (2001).

Birds (scavenger guild morphologies): *Hertel and Lehman* (1998).

Mammals (predatory guild morphologies): *Van Valkenburgh* (1985, 1988).

Mammals (primate facial morphologies): *Richtsmeier and Lele* (1993).

SKELETON SPACE

All marine or terrestrial animals that possess skeletons, internal or external: *Thomas and Reif* (1993), *Thomas et al.* (2000), *Thomas* (2005).

back in time the founder of theoretical morphology, Dave Raup, used analogue computers to produce the simulation graphics on oscilloscope screens! At a conference on 'Computational Approaches to Theoretical Morphology', called by the Santa Fe Institute in November of 2000, the computer scientist Przemyslaw Prusinkiewicz noted that Dave Raup's early computer simulations of growth in molluscs (Raup, 1961) used computer graphics *two years before* 'computer graphics' had been established as a field (Przemyslaw Prusinkiewicz, quoted in McGhee, 2001a) in that Foley and Van Dam (1982, p. 18) state that the 'beginnings of modern interactive computer graphics are found in Ivan Sutherland's seminal Ph.D. work on the Sketchpad system (1963)'.

Several published sources of source code for computer programs used in theoretical morphological simulations of organic form are given in Table 9.2 (see also the 'problem-solving environment' approach to computer simulation of Merks *et al.*, 2006). Unfortunately, many scientific journal editors are reluctant to publish computer source code, as the per-page costs for printing those journals is expensive, thus the list given in Table 9.2 is not as long as I would like it to be. Things are changing, however, and with the advent of electronic publishing and personal Websites on the Internet, much more information will become easily available in the future.

What more is to be done?

The Harvard palaeontologist Stephen Jay Gould, the twentieth century's most noted essayist on the wonders of life that evolution has

Table 9.2. *Selected published sources of source code for computer programs used in theoretical morphological simulations of organic form*

BASIC programs for accretionary growth systems, useful for generating hypothetical shell forms in gastropods, cephalopods, brachiopods, and bivalves: *Savazzi (1985), Okamoto (1988), Swan (1999), Ubukata (2000).*

BASIC programs for branching growth systems, useful for generating hypothetical tree forms and bryozoan forms: *Swan (1999), Raup, McGhee, and McKinney (2006).*

BASIC programs for laminar growth systems, useful for generating hypothetical stromatoporoid and stromatolite forms: *Swan (1999).*

C/C++ programs for accretionary growth systems, useful for generating hypothetical shell forms in gastropods, cephalopods, brachiopods, and bivalves: *Savazzi (1990, 1993).*

Mathematica programs for the computer simulation of a broad spectrum of biological form, from animals to plants: *Wolfram (2002).*

produced, wrote of the discipline of theoretical morphology: 'I believe that the question of defining morphospaces and mapping their differential filling through time is so vital to our understanding of life's history …' (Gould, 1991, p. 422). That defining and mapping task has begun, yet we are still in early days yet, and the numbers of actual organisms that have been studied by using the analytical techniques of theoretical morphology are still relatively small (Table 9.1).

The philosopher of science James Maclaurin has asked the rhetorical question 'What shall we do with theoretical morphology?' and has compiled the following list of observations and suggestions of future research programmes in the discipline:

(1) Theoretical morphology has already led to a better undestanding of growth in living organisms. This is the fundamental purpose of that branch of theoretical morphology, which undertakes the modelling of organic morphogenesis.
(2) A second key use of theoretical morphology is in the formulation and testing of adaptationist hypotheses [… the relationship between the current function of a trait and its selective history].
(3) [T]heoretical morphology's contribution to the investigation of evolutionary trends in morphospace.
(4) [T]heoretical morphology might one day shed light on the "great Cambrian disparity debate". One of Stephen Jay Gould's most famous claims is his attack on the idea that natural selection produces a cone of increasing diversity. In a succession of works, Gould (1989, 1991, 1993, 1995) advanced the thesis that biological disparity is, in fact, lower now than it was directly after the Cambrian explosion. Debate has raged ever since, as to how we might measure disparity in an effort to test the original claim.
(5) Theoretical morphology might aid in making operational the idea of the adaptive landscape.
(6) Theoretical morphology might allow us to better understand the nature of biodiversity and thus aid in the performance of conservation biology.
(7) Theoretical morphology might solve a problem identified by Sally Ferguson (2002) concerning evolutionary explanation. She notes that we think of traits as standing in need of evolutionary

explanation because of their complex functionality. However, if the function referred to can only be spelled out in teleological terms, then the whole idea of providing an evolutionary explanation begins to look circular. Using theoretical morphology may provide a means of avoiding this problem (because it allows us to focus upon the explanation of pattern rather than process).

(8) Finally, theoretical morphology might allow us to sort life into the actual, the non-actual and the impossible and, thereby, it might help us to better explore the fundamental constraints on living systems (Maclaurin, 2003, pp. 470–471).

He further makes the philosophical distinction between what he terms a *partial theoretical morphospace*, which is 'a hypothetical space with a limited number of dimensions that can be represented graphically and used to evaluate hypotheses about the histories of particular traits within particular lineages' (Maclaurin, 2003, p. 471), and the *total theoretical morphospace*, which is '... the totality of morphospace. It has many (perhaps uncountably many) dimensions and houses within it all possible biological form ... there are many partial theoretical morphospaces but only one total theoretical morphospace' (Maclaurin, 2003, pp. 471–472). Having drawn this distinction, he suggests that research programmes 1, 2, 3 and 7 are clearly in the domain of partial morphospace analyses, research programmes 4 and 8 are more likely in the domain of total morphospace analysis, and that research programmes 5 and 6 could be investigated by models of either type.

The *partial* versus *total* theoretical morphospace distinction of Maclaurin (2003) speaks directly to an objection raised to theoretical morphospace analyses of the 'great Cambrian disparity debate' by Hutchinson (1999), who posed the question 'but which morphospace to use?' of such analyses. He maintained that the amount and extent of biodisparity one can measure in morphospace is itself a function of the dimensionality of the morphospace, hence different morphospaces will reveal different patterns of disparity, and thus theoretical morphospace analyses can never definitively prove or disprove whether biological disparity was greater in the Cambrian than at present. Maclaurin's philosophical conceptualization reveals that Hutchinson's objection is true with regard to partial theoretical morphospaces, but false with regard to total theoretical morphospace. Thus, at least conceptually, the question can be addressed by the analytical techniques of theoretical morphology, as urged by Gould (1991).

The re-emergence of the science of morphology

Eble (2000) argues that a central concept of theoretical morphology, that existent organic forms are only a subset of the set of theoretically possible forms, can be traced back to Johann Wolfgang von Goethe. Goethe is indeed generally credited with being the founder of the science of morphology, in that he coined the term *Morphologie* itself and was very interested in the comparative anatomy of animals and plants. His work in the late 1700s and early 1800s inspired the next generation of morphologists. These include the great comparative anatomists Georges Cuvier, twenty years younger than Goethe, who founded the science of vertebrate palaeontology, and Richard Owen, contemporary of Charles Darwin and founder of the study of the 'terrible reptiles', the dinosaurs. Darwin himself, founder of the study of evolution through the process of natural selection, was very interested in comparative animal and plant morphologies. His famous insights into morphological variation as the raw material of evolution are just that, morphological. In *On the Origin of Species* itself, Darwin (1859, p. 434) wrote of comparative morphology:

> This is the most interesting department of natural history, and may be said to be its very soul. What can be more curious than that the hand of a man, formed for grasping, that of a mole for digging, the leg of the horse, the paddle of the porpoise, and the wing of the bat, should all be constructed on the same pattern, and should include the same bones, in the same relative position?

However, with the publication of Darwin's theory of natural selection, research in morphology began to shift from the analysis of the spectrum of existent form, and possible theoretical limits on that spectrum, to the analysis of shifting morphological frequencies within populations of organisms. With Gregor Mendel's discovery of genetics, the discovey that morphological inheritance was quantal and discrete, that trend accelerated. The very word *morphology* itself began to be replaced by *phenotype*. Research into phenotypic variation became increasingly reduced to research into genotypic variation.

In the 1930s, the role of theory blossomed in the study of evolution, yet it was theoretical population genetics and not theoretical morphology. The founders of the Neo-Darwinian synthesis, R. A. Fisher,

J. B. S. Haldane, and Sewall Wright, were all geneticists. The study of evolution began to be viewed as the study of genetics. Darwin's functional morphological understanding of the word *fitness* itself was replaced by a genetic definition of differential changes in gene frequencies.

The theoretical approaches to the analysis of morphology advocated by Bateson (1896), Cook (1914), Russell (1916) and Thompson (1917) were, as noted by Keller (2002), lauded but largely ignored. Thompson's theoretical mathematics did generate enough interest to spur him to write a new, and greatly expanded, version of his original book (Thompson, 1942), yet even in 1942 he writes:

> In the morphology of living things the use of mathematical methods and symbols has made slow progress; and there are various reasons for this failure to employ a method whose advantages are so obvious in the investigation of other physical forms. To begin with, there would seem to be a psychological reason, lying in the fact that the student of living things is by nature and training an observer of concrete objects and phenomena and the habit of mind which he possesses and cultivates is alien to that of the theoretical mathematician. But this is by no means the only reason; for in the kindred subject of mineralogy, for instance, crystals were still treated in the days of Linnaeus as wholly within the province of the naturalist, and were described by him after the simple methods in use for animals and plants: but as soon as Haüy shewed the application of mathematics to the description and classification of crystals, his methods were immediately adopted and a new science came into being.
>
> *(Thompson, 1942, p. 1028)*

But, although D'Arcy Thompson tried twice to show the power of the application of mathematics to the analysis of morphology, no new science came into being.

The importance of morphology in the study of evolution re-emerged with the publication of *Tempo and Mode in Evolution* (1944) and *The Major Features of Evolution* (1953) by the palaeontologist George Gaylord Simpson. Simpson was one of the architects, along with the biologist Ernst Mayr and the geneticist Theodosius Dobzhansky, of what came to be known as the Modern Synthesis of evolutionary theory. Intriguingly, it was the earlier fitness landscape concept of the Neo-Darwinian geneticist Sewall Wright that inspired Simpson to propose a new synthesis of population genetics and palaeontological approaches to the study of evolution. Simpson made the conceptual jump from the fitness landscape of genotypes to the adaptive landscape of morphologies, and using 'this phenotypic landscape, Simpson illustrated the concepts of phenotypic variation, selection, immediate responses to selection, long-term evolutionary trends, speciation and adaptive radiation. No visualization before or since 1944 has been so successful in integrating the major issues and themes in phenotypic evolution' (Arnold, Pfrender and Jones, 2001, p. 9).

Following the publication of Simpson's books, the study of morphological evolution experienced a resurgence in palaeontology and biology. In biology, however, the study of morphology was quickly eclipsed by the discovery of the structure of DNA by James Watson and Francis Crick in 1953, and the subsequent decoding of the coding mechanism of life itself. The discipline of molecular biology was born:

> To make matters worse for morphology, the research program for molecular evolution is conceptually simple, involving the comparative analysis of well-defined molecular units (i.e. bases, amino acids). All that is needed for comparison are some assumptions about the tempo in which the units mutate... Morphological information, on the other hand, is not that simple. Shape is an elusive concept...
>
> *(Rasskin-Gutman, 2003, p. 305)*

By the mid-1960s the DNA coding for all the amino acids had been worked out, and in the past four decades the discipline of molecular biology has experienced explosive growth in the number of its practitioners.

Within palaeontology the discipline of theoretical morphology was born during this same interval of time. The final key event for the actual utilization of the adaptive landscape concept as an analytical tool was the invention of the digital computer. By the mid-1960s digital computers were to be found on all major university campuses, and Dave Raup began his early computer simulation studies of mollusc shell form. He first was interested in the theoretical mathematical characterization of morphogenesis, inspired by D'Arcy Thompson's theory of transformations approach to comparative morphology. He soon discovered that the computer allowed him to explore morphological transformations that had never occurred in nature, to create nonexistent form. The concept of the theoretical morphospace was soon to follow, and the realization of its potential usefulness in the analysis of the adaptive significance of form: 'In studying the functional significance of the coiled shell, it is important to be able to analyze the types that do not occur in nature as well as those represented by actual species. Both digital and analog computers are useful in constructing accurate pictures of the types that do not occur' (Raup and Michelson, 1965, p. 1294). The science of morphology has finally entered into a new theoretical, rigorous and mathematical stage.

Within biology the realization slowly has been spreading in the past two decades that on a 'fundamental level, molecular evolution and morphological evolution are two quite different processes: the end products of both are linked and expressed by the nonlinear events that occur during the development of an organism' (Rasskin-Gutman, 2003, p. 306). The discipline of evolutionary development, or *Evo-Devo* to its practitioners, has been born, as well as a number of new scientific journals devoted exclusively to the empirical and theoretical study of processes that generate explicitly morphological patterns. In a review of the recently published evo-devo book *The Development of Animal Form: Ontogeny, Morphology, and Evolution* (2003) by Alessandro Minelli, Meyer (2003, p. 255) writes of the 'intellectual journey that every generation of self-respecting biologists has travelled since Ernst Haeckel, Karl Ernst, Ritter von Baer, Georges Cuvier and Johann Wolfgang von Goethe before them … Selection can only act on things that are developmentally possible. In other words, developmental mechanisms constrain evolutionary possibilities … But how do developmental mechanisms themselves change during evolution, and how does evolution in turn effect development?' Such questions are firmly in the realm of comparative morphology (Müller and Newman, 2005; Jablonski, 2005).

D'Arcy Thompson's vision of a theoretical mathematical approach to the science of morphology is now a reality, and a new science has come into being. Sewall Wright's and George Gaylord Simpson's vision of analysing evolution on an adaptive landscape is no longer simply a conceptual model, a heuristic tool, but is now an analytical technique for the morphologist to use. After decades of being sidelined in the wings, the science of morphology has returned to the stage in the analysis of the evolution of life (Müller and Newman, 2005). The morphologist Carole Hickman (1993, p. 170) writes of the future:

> The ultimate triumph of theoretical morphology would be an understanding of biological diversity, framed in terms of the boundaries between the possible and the actual and the possible and the impossible. It should integrate across all levels of structure, from organic molecules to entire and seemingly complex functioning organisms, where as yet undiscovered laws of structural consonance may exist.

From microevolutionary processes at the molecular level to macroevolutionary processes in geological time, the analysis of these processes 'could be tackled within the framework of the morphospace of connections by analyzing the dynamics involved in the establishment of boundaries during embryonic development and by looking at macroevolutionary boundary patterns. This is a task that almost certainly would have intrigued Darwin, just as he would have been fascinated with the discovery of the genetic code' (Rasskin-Gutman, 2003, p. 321).

References

Ackerly, S. C. (1989). Kinematics of accretionary shell growth, with examples from brachiopods and molluscs. *Paleobiology*, **15**, 147–164.

Ackerly, S. C. (1992). The structure of ontogenetic variation in the shell of *Pecten*. *Palaeontology*, **35**, 847–867.

Alberch, P. (1982). Developmental constraints in evolutionary processes. In *Evolution and Development*, ed. J. T. Bonner, pp. 313–332. Berlin: Springer Verlag.

Alberch, P. (1989). The logic of monsters: evidence for internal constraint in development and evolution. *Geobios, mémoire spécial*, **12**, 21–57.

Alberch, P. (1991). From genes to phenotype: dynamical systems and evolvability. *Genetica*, **84**, 5–11.

Alberch, P. and Gale, E. A. (1985). A developmental analysis of an evolutionary trend: digital reduction in amphibians. *Evolution*, **39**, 8–23.

Alberch, P., Gould, S. J., Oster, G. F., and Wake, D. B. (1979). Size and shape in ontogeny and phylogeny. *Paleobiology*, **5**, 296–317.

Antonovics, J. and van Tienderen, P. H. (1991). Ontoecogenophyloconstraints? The chaos of constraint terminology. *Trends in Ecology and Evolution*, **6**, 166–168.

Arnold, S. J., Pfrender, M. E., and Jones, A. G. (2001). The adaptive landscape as a conceptual bridge between micro- and macroevolution. *Genetica*, **112–113**, 9–32.

Bambach, R. K., Knoll, A. H., and Wang, S. C. (2004). Origination, extinction, and mass depletions in marine diversity. *Paleobiology*, **30**, 522–542.

Bateson, W. (1896). *Materials for the Study of Variation*. Baltimore: Johns Hopkins University Press.

Bayer, U. and McGhee, G. R. Jr. (1984). Iterative evolution of Middle Jurassic ammonite faunas. *Lethaia*, **17**, 1–16.

Bayer, U. and McGhee, G. R. Jr. (1985). Evolution in marginal epicontinental basins: the role of phylogenetic and ecological factors (ammonite replacements in the German Lower and Middle Jurassic). In *Sedimentary and Evolutionary Cycles*, eds. U. Bayer and A. Seilacher, pp. 164–220. Berlin: Springer Verlag.

Berger, W. H. (1969). Planktonic foraminifera: basic morphology and ecologic implications. *Journal of Paleontology*, **43**, 1369–1383.

Blomberg, S. P. and Garland, T. (2002). Tempo and mode in evolution: phylogenetic inertia, adaptation and comparative methods. *Journal of Evolutionary Biology*, **15**, 899–910.

Bonner, J. T. (1982). *Evolution and Development: Report of the Dahlem Workshop on Evolution and Development, Berlin 1981*. Berlin: Springer Verlag.

Bookstein, F. L. (1977). The study of shape transformation after D'Arcy Thompson. *Mathematical Biosciences*, **43**, 177–219.

Bookstein, F. L. (1997). *Morphometric Tools for Landmark Data: Geometry and Biology*. Cambridge: Cambridge University Press.

Brasier, M. D. (1980). *Microfossils*. London: George Allen and Unwin.

Cain, A. J. (1977). Variation in the spire index of some coiled gastropod shells, and its evolutionary significance. *Philosophical Transactions of the Royal Society of London (B: biological sciences)*, **277**, 377–428.

Callebaut, W. and Rasskin-Gutman, D. (2005). *Modularity: Understanding the Development and Evolution of Natural Complex Systems*. Cambridge (MA): Vienna Series in Theoretical Biology, Massachusetts Institute of Technology Press.

Cattin, M.-F., Bersier, L.-F., Banašek-Richter, C., Baltensperger, R., and Gabriel, J.-P. (2004). Phylogenetic constraints and adaptation explain food-web structure. *Nature*, **427**, 835–839.

Chamberlain, J. A. Jr. (1976). Flow patterns and drag coefficients of cephalopod shells. *Palaeontology*, **19**, 539–563.

Chamberlain, J. A. Jr. (1981). Hydromechanical design of fossil cephalopods. *Systematics Association Special Volume*, **18**, 289–336.

Checa, A. G., Okamoto, T., and Keupp, H. (2002). Abnormalities as natural experiments: a morphologic model for coiling regulation in planispiral ammonites. *Paleobiology*, **28**, 127–138.

Cheetham, A. H. and Hayek, L. C. (1983). Geometric consequences of branching growth in adeoniform Bryozoa. *Paleobiology*, **9**, 240–260.

Ciampaglio, C. N. (2002). Determining the role that ecological and developmental constraints play in controlling disparity: examples from the crinoid and blastozoan fossil record. *Evolution and Development*, **4**, 170–188.

Conway Morris, S. (1998). *The Crucible of Creation: The Burgess Shale and the Rise of Animals*. Oxford: Oxford University Press.

Conway Morris, S. (2003). *Life's Solution: Inevitable Humans in a Lonely Universe*. Cambridge: Cambridge University Press.

Cook, T. A. (1914). *The Curves of Life*. London: Constable and Company.

Cortie, M. B. (1989). Models for mollusc shape. *South African Journal of Science*, **85**, 454–460.

Cubo, J. (2004). Pattern and process in constructional morphology. *Evolution and Development*, **6**, 131–133.

Darwin, C. (1859). *On the Origin of Species by Means of Natural Selection, or the Preservation of Favoured Races in the Struggle for Life*. London: John Murray.

Davoli, F. and Russo, F. (1974). Una metodologia paleontometrica basata sul modello di Raup:verifica sperimentale su rappresentanti follili del gen. *Subula* Schumacher. *Bollettino della Società Paleontologica Italiana*, **13**, 108–121.

Dawkins, R. (1996). *Climbing Mount Improbable*. New York: W. W. Norton and Co.

Dennett, D. C. (1996). *Darwin's Dangerous Idea*. New York: Simon and Schuster.
Dobzhanski, T. (1970). *Genetics of the Evolutionary Process*. New York: Columbia University Press.
Dommergues, J.-L., Laurin, B., and Meister, C. (1996). Evolution of ammonoid morphospace during the Early Jurassic radiation. *Paleobiology*, **22**, 219–240.
Eble, G. J. (2000). Theoretical morphology: state of the art. *Paleobiology*, **26**, 520–528.
Eble, G. J. (2003). Developmental morphospaces and evolution. In *Evolutionary Dynamics:Exploring the Interplay of Selection, Accident, Neutrality, and Function*, eds. J. P. Crutchfield and P. Schuster, pp. 33–63. Oxford: Oxford University Press.
Ellers, O. (1993). A mechanical model of growth in regular sea urchins: predictions of shape and a developmental morphospace. *Proceedings of the Royal Society of London*, **B254**, 123–129.
Ellison, A. M. and Niklas, K. J. (1988). Branching patterns of *Salicornia europaea* (Chenopodiaceae) at different successional stages: a comparison of theoretical and real plants. *American Journal of Botany*, **75**, 501–512.
Ferguson, S. A. (2002). Methodology in evolutionary psychology. *Biology and Philosophy*, **17**, 635–650.
Fisher, R. A. (1941). Average excess and average effect of a gene substitution. *Annals of Eugenics*, **11**, 53–63.
Foley, J. and van Dam, A. (1982). *Fundamentals of Interactive Computer Graphics*. Reading, MA: Addison Wesley.
Fortey, R. A. (1983). Geometric constraints in the construction of graptolite stipes. *Paleobiology*, **9**, 116–125.
Funk and Wagnall (1963). *Standard College Dictionary*. New York: Harcourt, Brace and World, Inc.
Futuyma, D. (1998). *Evolutionary Biology*. Sunderland (MA): Sinauer Associates, Inc.
Gärdenfors, P. (2000). *Conceptual Spaces: the Geometry of Thought*. Cambridge, MA: MIT Press.
Gavrilets, S. (1997). Evolution and speciation on holey adaptive landscapes. *Trends in Ecology and Evolution*, **12**, 307–312.
Gavrilets, S. (1999). A dynamical theory of speciation on holey adaptive landscapes. *American Naturalist*, **154**, 1–22.
Gavrilets, S. (2003). Evolution and speciation in a hyperspace: the roles of neutrality, selection, mutation, and random drift. In *Evolutionary Dynamics: Exploring the Interplay of Selection, Accident, Neutrality, and Function*, eds. J. P. Crutchfield and P. Schuster, pp. 135–162. Oxford: Oxford University Press.
Gavrilets, S. and Gravner, J. (1997). Percolation on the fitness hypercube and the evolution of reproductive isolation. *Journal of Theoretical Biology*, **184**, 51–64.
Goodwin, B. C. (1963). *Temporal Organization in Cells*. London: Academic Press.
Gould, S. J. (1976). D'Arcy Thompson and the science of form. In *Topics in the Philosophy of Biology*, eds. M. Grene and E. Mendelsohn, pp. 66–97. Dortrecht: D. Reidel.

Gould, S. J. (1989). *Wonderful Life*. New York: W. W. Norton.
Gould, S. J. (1991). The disparity of the Burgess Shale arthropod fauna and the limits of cladistic analysis: why we must strive to quantify morphospace. *Paleobiology*, **17**, 411–423.
Gould, S. J. (1993). How to analyze Burgess Shale disparity – a reply to Ridley. *Paleobiology*, **19**, 522–523.
Gould, S. J. (1995). A task for paleobiology at the threshold of majority. *Paleobiology*, **21**, 1–14.
Hammer, Ø. and Bucher, H. (2005). Models for the morphogenesis of the molluscan shell. *Lethaia*, **38**, 111–122.
Hertel, F. and Lehman, N. (1998). A randomized nearest-neighbor approach for assessment of character displacement: the vulture guild as a model. *Journal of Theoretical Biology*, **190**, 51–61.
Hickman, C. S. (1993). Theoretical design space: a new paradigm for the analysis of structural diversity. *Neues Jahrbuch für Geologie und Paläontologie, Abhandlungen*, **190**, 169–182.
Honda, H. and Fisher, J. B. (1978). Tree branch angle: maximizing effective leaf area. *Science*, **199**, 888–890.
Hutchinson, J. M. C. (1999). But which morphospace to use? *Trends in Ecology and Evolution*, **14**, 414.
Jablonski, D. (2005). Evolutionary innovations in the fossil record: the intersection of ecology, development, and macroevolution. *Journal of Experimental Zoology Part B (Molecular and Developmental Evolution)*, **304B**, 504–519.
Kauffman, S. A. (1993). *The Origins of Order*. Oxford: Oxford University Press.
Kauffman, S. A. (1995). *At Home in the Universe*. Oxford: Oxford University Press.
Keller, E. F. (2002). *Making Sense of Life: Explaining Biological Development with Models, Metaphors, and Machines*. Cambridge, MA: Harvard University Press.
Kendrick, D. C. (2007). Theoretical morphology of crinoid calyxes. *Paleobiology*, in press.
Kershaw, S. and Riding, R. (1978). Parameterization of stromatoporoid shape. *Lethaia*, **11**, 233–242.
Kohn, A. J. and Riggs, A. C. (1975). Morphometry of the *Conus* shell. *Systematic Zoology*, **24**, 346–359.
Korn, D. (2000). Morphospace occupation of ammonoids over the Devonian–Carboniferous boundary. *Paläontologische Zeitschrift*, **74**, 247–257.
Kuhn-Schnyder, E. and Rieber, H. (1986). *Handbook of Paleozoology*. Baltimore: Johns Hopkins University Press.
McCartney, K. and Loper, D. E. (1989). Optimized skeletal morphologies of silicoflagellate genera *Dictyocha* and *Distephanus*. *Paleobiology*, **15**, 283–298.
McCartney, K. and Loper, D. E. (1992). Optimal models of skeletal morphology for the silicoflagellate genus *Corbisema*. *Micropaleontology*, **38**, 87–93.
McGhee, G. R. Jr. (1980a). Shell form in the biconvex articulate Brachiopoda: a geometric analysis. *Paleobiology*, **6**, 57–76.

McGhee, G. R. Jr. (1980b). Shell geometry and stability strategies in the biconvex Brachiopoda. *Neues Jahrbuch für Geologie und Paläontologie, Monatshefte*, **1980**(3), 155–184.

McGhee, G. R. Jr. (1988). The Late Devonian extinction event: evidence for abrupt ecosystem collapse. *Paleobiology*, **14**, 250–257.

McGhee, G. R. Jr. (1991). Theoretical morphology: the concept and its applications. In *Analytical Paleobiology*, eds. N. L. Gilinsky and P. W. Signor, pp. 87–102. Short Courses in Paleontology No. 4, the Paleontological Society and the Univeristy of Tennessee, Knoxville.

McGhee, G. R. Jr. (1995). Geometry of evolution in the biconvex Brachiopoda: morphological effects of mass extinction. *Neues Jahrbuch für Geologie und Paläontologie, Abhandlungen*, **197**, 357–382.

McGhee, G. R. Jr. (1999). *Theoretical Morphology: the Concept and Its Applications*. New York: Columbia University Press.

McGhee, G. R. Jr. (2001a). Exploring the spectrum of existent, nonexistent and impossible biological form. *Trends in Ecology and Evolution*, **16**, 172–173.

McGhee, G. R. Jr. (2001b). The question of spiral axes and brachiopod shell growth: a comparison of morphometric techniques. *Paleobiology*, **27**, 716–723.

McGhee, G. R. Jr. (2006). Exploring the spectrum of existent, nonexistent, and impossible biological form: a research program. In *Modeling Biology: Structures, Behaviors, Evolution*, eds. M. Laubichler and G. B. Müller, pp. in press. Cambridge (MA): Vienna Series in Theoretical Biology, Massachusetts Institute of Technology Press.

McGhee, G. R. Jr. and McKinney, F. K. (2000). A theoretical morphologic analysis of convergently evolved erect helical colony form in the Bryozoa. *Paleobiology*, **26**, 556–577.

McGhee, G. R. Jr. and McKinney, F. K. (2002). A theoretical morphologic analysis of ecomorphologic variation in *Archimedes* helical colony form. *Palaios*, **17**, 556–570.

McGhee, G. R. Jr. and Starcher, R. W. (2006). Geometric models of lophophore shape and arrangement in extinct modular organisms. *Journal of Paleontology*, in press.

McGhee, G. R. Jr., Bayer, U., and Seilacher, A. (1991). Biological and evolutionary responses to transgressive-regressive cycles. In *Cycles and Events in Stratigraphy*, eds. G. Einsele, W. Ricken, and A. Seilacher, pp. 696–708. Berlin: Springer Verlag.

McGhee, G. R. Jr., Sheehan, P. M., Bottjer, D. J., and Droser, M. L. (2004). Ecological ranking of Phanerozoic biodiversity crises: ecological and taxonomic severities are decoupled. *Palaeogeography, Palaeoclimatology, Palaeoecology*, **211**, 289–297.

McGowan, A. J. (2004). The effect of the Permo-Triassic bottleneck on Triassic ammonoid morphological evolution. *Paleobiology*, **30**, 369–395.

McHenry, M. J. and Patek, S. N. (2004). The evolution of larval morphology and swimming performance in ascidians. *Evolution*, **58**, 1209–1224.

Mack, R. N. (2003). Phylogenetic constraint, absent life forms, and preadapted alien plants: a prescription for biological invasions. *International Journal of Plant Science*, **164**(3 Supplement), S185–S196.

McKinney, F. K. and McGhee, G. R. Jr. (2003). Evolution of erect helical colony form in the Bryozoa: phylogenetic, functional, and ecological factors. *Biological Journal of the Linnean Society*, **80**, 360–367.

McKinney, F. K. and McGhee, G. R. Jr. (2004). Erratum: Evolution of erect helical colony form in the Bryozoa. *Biological Journal of the Linnean Society*, **81**, 619–620.

McKinney, F. K. and Raup, D. M. (1982). A turn in the right direction: simulation of erect spiral growth in the bryozoans *Archimedes* and *Bugula*. *Paleobiology*, **8**, 101–112.

McKitrick, M. C. (1993). Phylogenetic constraint in evolutionary theory: has it any explanatory power? *Annual Review of Ecology and Systematics*, **24**, 307–330.

Maclaurin, J. (2003). The good, the bad and the impossible. *Biology and Philosophy*, **18**, 463–476.

Maynard Smith, J., Burian, R., Kauffman, S., Alberch, P., Campbell, J., Goodwin, B., Lande, R., Raup, D., and Wolpert, L. (1985). Developmental constraints and evolution. *Quarterly Review of Biology*, **60**, 265–287.

Merks, R. M. H., Hoekstra, A. G., Kaandorp, J. A., Sloot, P. M. A., and Hogeweg, P. (2006). Problem-solving environments for biological morphogenesis. *Computing in Science and Engineering*, **8**(1), 61–72.

Meyer, A. (2003). There and back again. *Nature*, **424**, 255.

Minelli, A. (2003). *The Development of Animal Form: Ontogeny, Morphology, and Evolution*. Cambridge: Cambridge University Press.

Müller, G. B. and Newman, S. A. (2003). *Origination of Organismal Form: Beyond the Gene in Deveopmental and Evolutionary Biology*. Cambridge (MA): Vienna Series in Theoretical Biology, Massachusetts Institute of Technology Press.

Müller, G. B. and Newman, S. A. (2005). The innovation triad: an evodevo agenda. *Journal of Experimental Zoology Part B (Molecular and Developmental Evolution)*, **304B**, 487–503.

Müller, G. B. and Wagner, G. P. (1991). Novelty in evolution: restructuring the concept. *Annual Review of Ecology and Systematics*, **22**, 229–256.

Newman, M. E. J. and Palmer, R. G. (2003). *Modeling Extinction*. Oxford: Oxford University Press.

Niklas, K. J. (1986). Computer-simulated plant evolution. *Scientific American*, **254** (March), 78–86.

Niklas, K. J. (1997a). Effects of hypothetical developmental barriers and abrupt environmental changes on adaptive walks in a computer-generated domain for early vascular land plants. *Paleobiology*, **23**, 63–76.

Niklas, K. J. (1997b). *The Evolutionary Biology of Plants*. Chicago: University of Chicago Press.

Niklas, K. J. (2004). Computer models of early land plant evolution. *Annual Review of Earth and Planetary Sciences*, **32**, 47–66.

Niklas, K. J. (2006). Optimization and early land plant evolution. In *Modeling Biology: Structures, Behavior, Evolution*, eds. M. Laubichler and G. B. Müller, in press. Cambridge (MA): Vienna Series in Theoretical Biology, Massachusetts Institute of Technology Press.

Niklas, K. J. and Kerchner, V. (1984). Mechanical and photosynthetic constraints on the evolution of plant shape. *Paleobiology*, **10**, 79–101.

Okamoto, T. (1988). Analysis of heteromorph ammonoids by differential geometry. *Palaeontology*, **31**, 35–52.

Olsen, E. C. and Miller, R. L. (1958). *Morphological Integration*. Chicago: University of Chicago Press.

Oster, G. F., Shubin, N., Murray, J. D., and Alberch, P. (1988). Evolution and morphogenetic rules: the shape of the vertebrate limb in ontogeny and phylogeny. *Evolution*, **42**, 862–884.

Popov, I. Y. (2002). "Periodical systems" in biology (a historical issue). *Verhandlungen zur Geschichte und Theorie der Biologie*, **9**, 55–68.

Rasskin-Gutman, D. (2003). Boundary constraints for the emergence of form. In *Origination of Organismal Form: Beyond the Gene in Developmental and Evolutionary Biology*, eds. G. B. Müller and S. A. Newman, pp. 305–322. Cambridge (MA): Vienna Series in Theoretical Biology, Massachusetts Institute of Technology Press.

Rasskin-Gutman, D. (2005). Modularity: jumping forms within morphospace. In *Modularity: Understanding the Development and Evolution of Natural Complex Systems*, eds. M. Callebaut and D. Rasskin-Gutman, pp. 207–219. Cambridge (MA): Vienna Series in Theoretical Biology, Massachusetts Institute of Technology Press.

Rasskin-Gutman, D. and Buscalioni, A. D. (2001). Theoretical morphology of the Archosaur (Reptilia: Diapsida) pelvic girdle. *Paleobiology*, **27**, 59–78.

Rasskin-Gutman, D. and Izpisúa-Belmonte, J. C. (2004). Theoretical morphology of developmental asymmetries. *BioEssays*, **26**, 405–412.

Raup, D. M. (1961). The geometry of coiling in gastropods. *Proceedings of the National Academy of Sciences (USA)*, **47**, 602–609.

Raup, D. M. (1966). Geometric analysis of shell coiling: general problems. *Journal of Paleontology*, **40**, 1178–1190.

Raup, D. M. (1967). Geometric analysis of shell coiling: coiling in ammonoids. *Journal of Paleontology*, **41**, 43–65.

Raup, D. M. and Michelson, A. (1965). Theoretical morphology of the coiled shell. *Science*, **147**, 1294–1295.

Raup, D. M., McGhee, G. R., Jr., and McKinney, F. K. (2006). Source code for theoretical morphologic simulation of helical colony form in the Bryozoa. *Palaeontologia Electronica*, **9**(2); http://palaeo-electronica.org/paleo/2006_2/helical/index.html.

Reif, W.-E. (1980). A model of morphogenetic processes in the dermal skeleton of elasmobranchs. *Neues Jahrbuch für Geologie und Paläontologie, Abhandlungen*, **159**, 339–359.

Rex, M. A. and Boss, K. J. (1976). Open coiling in recent gastropods. *Malacologia*, **15**, 289–297.

Rice, S. H. (1997). The analysis of ontogenetic trajectories: when a change in size or shape is not heterochrony. *Proceedings of the National Academy of Sciences USA*, **94**, 907–912.

Richtsmeier, J. T. and Lele, S. (1993). A coordinate-free approach to the analysis of growth patterns: models and theoretical considerations. *Biological Reviews*, **68**, 381–411.

Ridley, M. (1996). *Evolution.* Cambridge (MA): Blackwell Science.

Russell, E. S. (1916). *Form and Function: a Contribution to the History of Animal Morphology.* Chicago: University of Chicago Press (1982 Reprint).

Saunders, W. B. and Swan, A. R. H. (1984). Morphology and morphologic diversity of mid-Carboniferous (Namurian) ammonoids in time and space. *Paleobiology*, **10**, 195–228.

Saunders, W. B., Work, D. M., and Nikolaeva, S. V. (2004). The evolutionary history of shell geometry in Paleozoic ammonoids. *Paleobiology*, **30**, 19–43.

Savazzi, E. (1985). SHELLGEN: A BASIC program for the modeling of molluscan shell ontogeny and morphogenesis. *Computers and Geosciences*, **11**, 521–530.

Savazzi, E. (1987). Geometric and functional constraints on bivalve shell morphology. *Lethaia*, **20**, 293–306.

Savazzi, E. (1990). C programs for displaying shaded three-dimensional objects on a PC. *Computers and Geosciences*, **16**, 195–209.

Savazzi, E. (1993). C++ classes for theoretical shell morphology. *Computers and Geosciences*, **19**, 931–964.

Schindel, D. E. (1990). Unoccupied morphospace and the coiled geometry of gastropods: architectural constraint or geometric covariation? In *Causes of Evolution*, eds. R. A. Ross and W. D. Allmon, pp. 270–304. Chicago: University of Chicago Press.

Schwenk, K. (1995). A utilitarian approach to constraint. *Zoology*, **98**, 251–262.

Signes, M., Bijma, J., Hemleben, C., and Ott, R. (1993). A model for planktic foraminiferal shell growth. *Paleobiology*, **19**, 71–91.

Simpson, G. G. (1944). *Tempo and Mode in Evolution.* New York: Columbia University Press.

Simpson, G. G. (1953). *The Major Features of Evolution.* New York: Columbia University Press.

Snoad, N. and Nilsson, M. (2003). Quasispecies evolution on dynamic fitness landscapes. In *Evolutionary Dynamics: Exploring the Interplay of Selection, Accident, Neutrality, and Function*, eds. J. P. Crutchfield and P. Schuster, pp. 273–289. Oxford: Oxford University Press.

Solé, R. (2002). Modelling macroevolutionary patterns: an ecological perspective. In *Biological Evolution and Statistical Physics*, eds. M. Lässig and A. Valleriani, pp. 312–337. Berlin: Springer Verlag.

Solé, R. and Goodwin, B. (2000). *Signs of Life: How Complexity Pervades Biology.* New York: Basic Books.

Stadler, P. F. (2002). Fitness landscapes. In *Biological Evolution and Statistical Physics*, eds. M. Lässig and A. Valleriani, pp. 183–204. Berlin: Springer Verlag.

Stadler, B. M. R. and Stadler, P. F. (2004). The topology of evolutionary biology. In *Modeling in Molecular Biology*, eds. G. Ciobanu and G. Rozenberg, pp. 267–286. Berlin: Springer Verlag.

Stadler, B. M. R., Stadler, P. F., Wagner, G. P., and Fontana, W. (2001). The topology of the possible: formal spaces underlying patterns of evolutionary change. *Journal of Theoretical Biology*, **213**, 241–274.

Starcher, R. W. and McGhee, G. R. Jr. (2000). Fenestrate theoretical morphology: geometric constraints on lophophore shape and arrangement in extinct Bryozoa. *Paleobiology*, **26**, 116–136.

Starcher, R. W. and McGhee, G. R. Jr. (2002). Theoretical morphology of modular organisms: geometric constraints of branch and dissepiment width and spacing in fenestrate bryozoans. *Neues Jahrbuch für Geologie und Paläontologie, Abhandlungen*, **223**, 79–122.

Starcher, R. W. and McGhee, G. R. Jr. (2003). Fenestrate graptolite theoretical morphology: geometric constraints on lophophore shape and arrangement in extinct hemichordates. *Journal of Paleontology*, **77**, 360–367.

Stone, J. R. (1996). Computer simulated shell shape and size variation in the Caribbean land snail genus *Cerion*: a test of geometrical constraints. *Evolution*, **50**, 341–347.

Stone, J. R. (1998). Ontogenic tracks and evolutionary vestiges in morphospace. *Biological Journal of the Linnean Society*, **64**, 223–238.

Stone, J. R. (1999). Using a mathematical model to test the null hypothesis of optimal shell construction by four marine gastropods. *Marine Biology*, **134**, 397–403.

Stone, J. R. (2002). Delayed prezygotic isolating mechanisms: evolution with a twist. *Proceedings of the Royal Society of London*, **269**, 861–865.

Stone, J. R. (2004). Nonoptimal shell forms as overlapping points in functional and theoretical morphospaces. *American Malacological Bulletin*, **18**, 129–134.

Strathmann, R. R. (1978). Progressive vacating of adaptive types during the Phanerozoic. *Evolution*, **32**, 907–914.

Streidter, G. F. (2003). Epigenesis and evolution of brains: from embryonic divisions to functional systems. In *Origination of Organismal Form: Beyond the Gene in Developmental and Evolutionary Biology*, eds. G. B. Müller and S. A. Newman, pp. 287–303. Cambridge (MA): Vienna Series in Theoretical Biology, Massachusetts Institute of Technology Press.

Swan, A. R. H. (1990). A computer simulation of evolution by natural selection. *Journal of the Geological Society of London*, **147**, 223–228.

Swan, A. R. H. (1999). Computer models of fossil morphology. In *Numerical Palaeobiology*, ed. D. A. T. Harper, pp. 157–179. London: John Wiley and Sons Ltd.

Swan, A. R. H. and Kershaw, S. (1994). A computer model for skeletal growth of stromatoporoids. *Palaeontology*, **37**, 409–423.

Thom, R. (1975). *Structural Stability and Morphogenesis: an Outline of a General Theory of Models*. Reading, MA: W. A. Benjamin, Inc.

Thomas, R. D. K. (2005). Hierarchial integration of modular structures in the evolution of animal skeletons. In *Modularity: Understanding the*

Development and Evolution of Natural Complex Systems, eds. M. Callebaut and D. Rasskin-Gutman, pp. 239–258. Cambridge (MA): Vienna Series in Theoretical Biology, Massachusetts Institute of Technology Press.

Thomas, R. D. K. and Reif, W.-E. (1993). The skeleton space: a finite set of organic designs. *Evolution*, **47**, 341–360.

Thomas, R. D. K., Shearman, R. M., and Stewart, G. W. (2000). Evolutionary exploitation of design options by the first animals with hard skeletons. *Science*, **288**, 1239–1242.

Thompson, D'A. W. (1917). *On Growth and Form*. Cambridge: Cambridge University Press.

Thompson, D'A. W. (1942). *On Growth and Form*. Cambridge: Cambridge University Press.

Tyszka, J. (2006). Morphospace of foraminiferal shells: results from the moving reference model. *Lethaia*, **39**, 1–12.

Tyszka, J. and Topa, P. (2005). A new approach to modeling of foraminiferal shells. *Paleobiology*, **31**, 522–537.

Ubukata, T. (2000). Theoretical morphology of hinge and shell form in Bivalvia: geometric constraints derived from space conflict between umbones. *Paleobiology*, **26**, 606–624.

Ubukata, T. (2001). Stacking increments: a new model and morphospace for the analysis of bivalve shell growth. *Historical Biology*, **15**, 303–321.

Ubukata, T. (2003a). A theoretical morphologic analysis of bivalve ligaments. *Paleobiology*, **29**, 369–380.

Ubukata, T. (2003b). Pattern of growth rate around aperture and shell form in Bivalvia: a theoretical morphological study. *Paleobiology*, **29**, 480–491.

Ubukata, T. (2005). Theoretical morphology of bivalve shell sculptures. *Paleobiology*, **31**, 643–655.

Van Valen, L. (1973). A new evolutionary theory. *Evolutionary Theory*, **1**, 1–30.

Van Valkenburgh, B. (1985). Locomotor diversity within past and present guilds of large predatory mammals. *Paleobiology*, **11**, 406–428.

Van Valkenburgh, B. (1988). Trophic diversity in past and present guilds of large predatory mammals. *Paleobiology*, **14**, 155–173.

Waddington, C. H. (1957). *The Strategy of the Genes: a Discussion of some Aspects of Theoretical Biology*. London: Allen and Unwin.

Waddington, C. H. (1975). *The Evolution of an Evolutionist*. Ithaca: Cornell University Press.

Wagner, G. P. (2001). What is the promise of developmental evolution? Part II: A causal explanation of evolutionary innovations may be impossible. *Journal of Experimental Zoology*, **291**, 305–309.

Wagner, G. P. and Altenberg, L. (1996). Complex adaptations and the evolution of evolvability. *Evolution*, **50**, 967–976.

Ward, P. (1980). Comparative shell shape distributions in Jurassic–Cretaceous ammonites and Jurassic-Tertiary nautilids. *Paleobiology*, **6**, 32–43.

Waters, J. A. (1977). Quantification of shape by use of Fourier analysis: the Mississippian blastoid genus *Pentremites*. *Paleobiology*, **3**, 288–299.

Williamson, P. G. (1981). Palaeontological documentation of speciation in Cenozoic molluscs from Turkana Basin. *Nature*, **293**, 437–443.

Wilson, E. O. and Bossert, W. H. (1971). *A Primer of Population Biology*. Sunderland: Sinauer.

Wolfram, S. (2002). *A New Kind of Science*. Champaign, IL: Wolfram Media, Inc.

Wray, G. A. (2002). Do convergent developmental mechanisms underlie convergent phenotypes? *Brain, Behavior and Evolution*, **59**, 327–336.

Wright, S. (1932). The roles of mutation, inbreeding, crossbreeding and selection in evolution. *Proceedings of the Sixth International Congress of Genetics*, **1**, 356–366.

Zwieniecki, M. A., Boyce, C. K., and Holbrook, N. M. (2004). Functional design space of single-veined leaves: role of tissue hydraulic properties in constraining leaf size and shape. *Annals of Botany*, **94**, 507–513.

Index